SUN'S INFLUENCE COSMIC RAYS AND WEATHER

BY
AJAY GHOSH

CONTENTS

CHAPTER 1 INTRODUCTION 1-29

    1.1   Cosmic Ray Genesis

    1.2   Discovery of Cosmic Rays

    1.3   Configuration of Cosmic Rays

    1.4   Range of Cosmic Ray's Energy

    1.5   Classification of Cosmic Rays

    1.6   Cascade Development of Cosmic Rays

    1.7   Lateral and Longitudinal Development of Charged Particles

    1.8   Cosmic Ray Modulation

CHAPTER 2 LITERATURE REVIEW 30-36

    2.1 Introduction

    2.2 Review of Literature

CHAPTER 3 EFFECT OF SOLAR ACTIVITY ON COSMIC RAY MODULATION 37-53

    3.1   Introduction

    3.2   Factors affecting Galactic Cosmic ray modulation

    3.3   Data Analysis

    3.4   Results and Discussion

    3.5   Conclusion

CHAPTER 4 VARIATIONS IN COSMIC RAYS WITH THE 11-YEARS SOLAR ACTIVITY 54-67

    4.1   Introduction

    4.2   Main factors Affecting variations in Cosmic rays

    4.3   Data Analysis

    4.4   Results and Discussion

    4.5   Conclusion

CHAPTER 5 STUDY OF EFFECT OF SOLAR ACTIVITY ON EARTH'ATMOSPHERIC WEATHER 68-101

    5.1   Introduction

    5.2   Coronal mass ejections

    5.3   F10.7 solar radio flux

    5.4   K-index

    5.5   Radio propagation

    5.6   The Ap-Index

    5.7   Disturbance storm time index

    5.8   Results and Discussion

    5.9   Conclusion

CHAPTER 6 WAVE ANALYSIS OF COSMIC RAYS ASSOCIATED WITH SOLAR-WIND AND GEOMAGNETIC INDEX (Ap) **102-114**

    6.1 Introduction

    6.2 Solar Wind

    6.3 Geomagnetic Storm

    6.4 Geomagnetic Storm effects

    6.5 Results

    6.6 Conclusions

CHAPTER 7 EFFECT OF SOLAR OUTPUT ON HIGH AND LOW AMPLITUDE WAVE TRAINS OF DIURNAL VARIATION OF COSMIC RAY INTENSITY **115-133**

    7.1 Introduction

    7.2 Factors affecting the impact of solar output on High and Low Amplitude Wave Trains of Diurnal Variation of Cosmic Ray Intensity

    7.3 Factors affecting the Solar Cycle

    7.4 High Amplitude Wave Trains of Cosmic Ray Intensity

    7.5 Factors affecting the High and Low Amplitude Wave CRI

    7.6 Data Analysis

7.7 Results and Discussion

7.8 Conclusion

CHAPTER 8   CONCLUSION **134-148**

CHAPTER 1
INTRODUCTION

1.1 COSMIC RAY GENESIS

Cosmic rays are high energy charged particles, which are originating in outer space. Velocity of Cosmic rays are similar to the speed of light and strike the Earth from all directions .They are atomic nuclei, ranging from the lightest to the heaviest elements in the periodic table .Cosmic rays also include high energy electrons ,positrons and other subatomic particles .Cosmic rays are also produced from solar energetic particles produced by sun and Cosmic particles are accelerated into the interstellar space. Cosmic rays are produced from supernovae, rotating neutron stars and black holes.

Cosmic rays have extremely high energies while some have relatively low energies .The shock waves formed in supernovae accelerate due to which high energy Cosmic rays having energy up-to 10^{14}eV are formed .High energy Cosmic rays approach the Earth's surface equally from all directions .The path of trajectory of Cosmic rays is spiral because of galactic magnetic fields .For low energy Cosmic rays (having energies below 10 GeV) the charged component of the Cosmic rays interact with the earth's magnetic field due to which the movement of Cosmic rays depend on the direction . The existence of Cosmic rays was detected by using simple experiments on electroscopes. Many workers flew ionization chambers in balloons for the detection and measurement of these penetrating radiations coming from outer space, and it was noticed that the remaining electrical conductivity remained in the chambers even at great distances from the earth's surface. In 1912-14, Victor F. Hess and Kolhorster showed that the ionization first decreased up to 700m and then increased with the altitude of the recording instrument. They deduced that the increase in ionization currents in the ionization chamber with height was due to the penetrating radiations coming from outer space. These radiations did not show a difference in intensity during day and night in non-solar genesis. Millikan and his colleagues took observations on mountain altitudes and snow-covered lakes at high altitudes

.They firmly manifested that the absorption of these radiations in water was similar to that produced in the atmospheric air. Millikan coined the name "Cosmic Rays" for these radiations. Several groups of Scientists in different countries including India got interested in the phenomenon of Cosmic Rays and started comprehensive investigation of Cosmic rays. Contributions of Bhabha ,Menon ,Sreekantan and Lal at T.I.F.R. Bombay are remarkable.Johnson and Street in 1933 proved that the intensity of radiations coming from eastward direction and that coming from westward direction are different. This phenomenon is called "East-West effect" and it confirmed that Cosmic rays are predominantly positively charged particles.

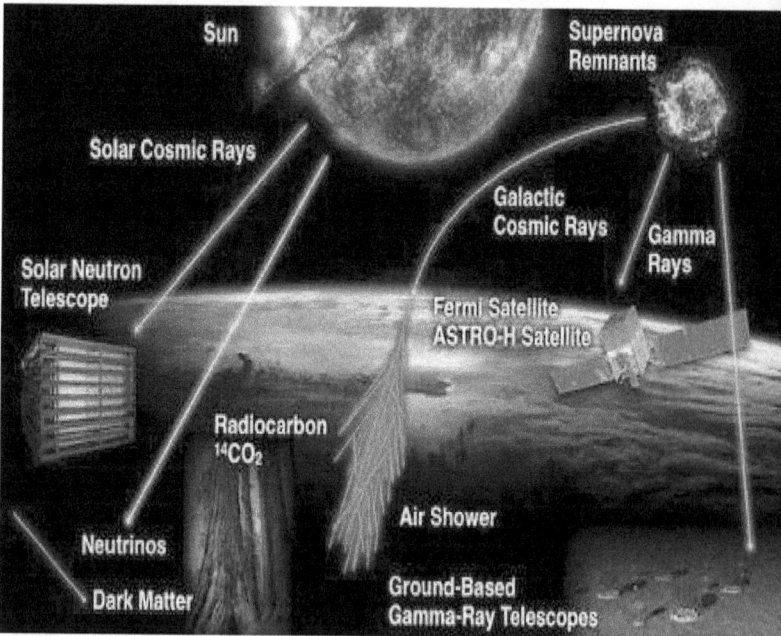

Fig 1.1 Earth and Cosmic Ray Interaction

1.2 Discovery of Cosmic Rays:

When Henry Coulomb noticed in 1785, the mystery of continuous and uncontrollable leakage of electric charge from a well insulated charged gold leaf electroscope remained unexplained but in 1900 ,H. Geital and J. Elster performed experiment on ionization of air and they observed some strange steady ionisation current in the perfectly shielded electrometers and this experiment opened avenue of new field of physics for the next generation scientists. T. Wulf ascended the Eiffel Tower in Paris taking ionization measuring device and observed that the ionisation decreased as he climbed to higher level. Gockel observed this ionisation at higher heights using balloons and concluded that the ionisation cannot be attributed to the radioactivity of the Earth. Austrian Physicist Prof. Victor F. Hess solved the issue unequivocally. On the historic day of August 7, 1912, Victor Hess along with an assistant went up in a balloon Gondola carrying three electroscopes and floated for several hours at altitudes ranging from 13,000 to 16,000 ft. and established that the radiation responsible for this obscure ionization is due to above and has an extra terrestrial origin. Through many large number of observations by a large number of workers it was established that the radiation is not only extra terrestrial but also extra-solar. Millikan confirmed this radiation and gave name in 1925 as "Cosmic Radiation". Prof. Victor Hess was awarded Nobel Prize for his discovery in the year 1936.The energy of cosmic ray distributes over a wide range from 1MeV to 10^{20} eV and possibly even beyond which is far from any artificial manmade accelerators. Such high energy particles can be created only by astrophysical phenomenon which are investigated by many scientists.

1.3 Configuration of Cosmic Rays

Cosmic rays can be broadly classified into two categories ,primary and secondary cosmic rays .The Primary Cosmic rays are produced through extra solar astronomical sources ,like galaxy clusters ,supernova remnants , binary stars ,black holes ,neutron stars etc.Primary Cosmic rays interact with celestial matter to form Secondary Cosmic rays.Sun emits Cosmic rays through a series of eruptions of intense high-energy radiations from the sun's surface due to sunspots which are of low intensity and are also responsible for the continuous production of a large number of unstable isotopes in the earth's atmosphere ,such as carbon – 14. It is due to Cosmic rays that the level of carbon-14 isotope in the atmosphere has remained constant since the past 100,000 years ,upto the inception of above-ground nuclear weapons testing in the prior 1950's .In archaeology ,Cosmic rays are an important tool as they interact with atmospheric nitrogen to form radiocarbon which is also used in radiocarbon dating. About 90% of all the incoming Cosmic Rays are protons, about 9% are helium nuclei (alpha particles) and about 1% are electrons. The remaining fraction is made up of the other heavier nuclei which are ample end products of star's nuclear synthesis. Secondary Cosmic rays consist of the other nuclei which are not abundant nuclear synthesis end products,products of the Big Bang primarily lithium ,beryllium and boron. These lighter nuclei are present in Cosmic rays in very large quantities (about 1:100 particles) than in solar atmospheres ,their abundance is about 10^{-7} that of helium.

When the heavy nuclei components of Primary Cosmic rays mainly the carbon and oxygen nuclei , collide with interstellar matter ,and they disintegrate into lighter nuclei i.e. ,lithium ,Beryllium and boron through a process called Cosmic Ray Spallation .The energy spectra of lithium, beryllium and boron falls off somewhat abruptly than carbon or oxygen, indicating that less Cosmic ray spallation occurs for the higher energy nuclei assumingly due to their escape from the galactic magnetic field. Cosmic rays consist of Scandium ,Titanium ,Vanadium and Manganese elements in large quantities for which Spallation is responsible. These elements are produced by collisions of iron and nickel nuclei with interstellar matter ; the local galactic magnetic field cannot accommodate particles with such high energies. Cosmic rays having energies above 10 GeV approach the earth's surface equally from all directions .Cosmic rays travel in spiral paths due to galactic magnetic fields ,and due to distribution of Cosmic ray sources .Hence Cosmic rays do not convey any information of their direction of origin.Cosmic rays having energy below 10 GeV have a directional dependence due to the interaction of the charged component of the Cosmic rays with the earth's magnetic field .When particles of Cosmic rays enter the earth's atmosphere and they collide with oxygen and nitrogen molecules and produces a cascade of lighter particles , called air-shower .Cosmic rays are accountable for yielding a large number of unstable isotopes in the earth's atmosphere, such as Carbon-14.

1.4 Range of Cosmic Ray's Energy

A cosmic ray is measured in MeV (Mega Electron Volts) or GeV (Giga Electron volts). Galactic Cosmic rays have energies ranging between 100 MeV - 10 GeV.Cosmic rays have been observed in the energy range 10^9 eV to over 10^{20} eV.Over this gamut, Cosmic rays follows a single power law ~ E^{-3}[8].The Energy spectrum seems to be a smooth curve over 10 decades of energy with a few distinguishable structures .Of these distinguishable structures ,the most important ones are the small ,abrupt changes in the local spectral index having power exponent ~3 just above 10^{15} eV and again just above 10^{18} eV.Cosmic rays having energies above 10^{18} eV are called Ultra-High Energy Cosmic Rays (UHECR).These are infinitesimal particles with a comprehensive amount of energy- about a joule or more.The energy contributed to the galaxy by Cosmic rays (about 1 eV per cm^3) is about equal to that contained in galactic magnetic fields.

Most galactic Cosmic rays acquire their energy from supernova explosions ,which occur approximately once every 50 years in our galaxy.

1.5 Classification of Cosmic Rays

Cosmic ray particles are classified into three main categories ,according to their origin and energy range :

1. Galactic Cosmic Rays
2. Anomalous Cosmic Rays
3. Solar Cosmic Rays

1.5.1 Galactic Cosmic Rays-

Galactic Cosmic rays (GCRs) come from outside the solar system but generally from within our Milky Way Galaxy.Galactic Cosmic rays are also called High Energy Cosmic Rays. The solar wind and solar flares consist of a ceaseless series of plasma ,loose protons and electrons .Heliosphere is the region in which the effect of solar wind and solar flare is felt .Since the solar wind is plasma ,it is an electrical conduction and it transmits a part of the sun's magnetic field .When GCR's move towards the Sun they confront the heliosphere and the magnetic field inside it and due to the structure of the magnetic field, the GCR's lose some of their energy.During high solar activity ,GCR's reach the earth as they lose some of their energy.Galactic Cosmic Rays have energies between 100 MeV and 10 GeV. GCR's are atomic nuclei from which all of the surrounding electrons have been stripped away during their high speed passage through the galaxy.Presumably they are accelerating since the last few million years ,traveled many times across the galaxy ,trapped by the galactic magnetic field. By supernova remnants ,GCR's have been accelerated to nearly the speed of light .On traveling through the very thin gas of celestial space ,GCR's interact and emit gamma rays ,which helps to know they pass through the Milky Way and other galaxies. Galactic Cosmic Rays are further classified into two categories :

(i) Primary Cosmic Rays
(ii) Secondary Cosmic Rays

Primary Cosmic Rays

The intensity of Cosmic rays outside the earth's atmosphere and away from the influence of geomagnetic field is called Primary Cosmic Rays. The flux of particles of Primary Cosmic Rays consist of positively charged particles which are the nuclei of normal elements found in the earth ,in stars and in interstellar space. The principal component is Hydrogen nuclei (Proton) about 93% and the next most important component is Helium nuclei (α-particles) about 6%. The heavy nuclei right upto Uranium (Z = 92) & possibly still heavier constitute about 1% including the very heavy ions also. In this context ,two book seem to be most probable :

The quasi-stationary background of particles at T≤15 MeV per nucleon is formed by galactic Cosmic Rays modulated by interplanetary magnetic fields .This book faces with some difficulties for explaining the rise in the spectrum with decreasing energy, because the modulation should have swept out low-energy particles more effectively .

The background is formed by particles accelerated inside the heliosphere boundary. Specially, particle's acceleration may occur in the region where the solar wind interacts with the interstellar medium or at stationary shocks which probably exist in this region.

Secondary Cosmic Rays-

When the Primary Cosmic Rays are incident at the top of our atmosphere interact with the oxygen and nitrogen air nuclei they produce new energetic particles called as Secondary Cosmic Rays. The products of Cosmic rays (secondary CR) fall broadly into three groups :

The soft or electromagnetic component

The hard or meson component

Nucleonic component

The soft or electromagnetic component is produced from the uncharged π^0 quickly into pairs of high energy gamma rays. Each Υ ray produces eventually an electron – positron pair (known as pair production) leading to the development of a cascade process through the atmosphere composed of high energy electrons, positrons and electromagnetic radiation (Υ –rays). This is called the "Soft" component of Cosmic rays because of its strong absorption by matter.

The hard or meson component begins with the charged pions $\pi^{\pm}$ which have life time of 10^{-8} sec. & decay into charged muons $\mu^{\pm}$ which then decay into $e^{\pm}$. Some muons are captured by other nuclei which sets a competition between decay and capture producing on one side decay products μ – mesons (or muons) of great penetrating power having relatively long lifetime of 2×10^{-6} sec. & which are not readily captured by nuclei. The mesons constitute most of the cosmic radiation observed at sea level & underground because of their ability to penetrate matter. In the Nucleonic component, the high energy neutrons & protons ejected from the primary interactions proceed through the atmosphere producing secondary & tertiary interactions which wholly or partially evaporate the nuclei involved. These interactions yield lower energy protons, neutrons & α – particles with initial energies of the order 2 to 30 MeV. These low energy protons and α – particles gets lost in ionization process and the neutrons are rapidly slowed down by collision processes in the atmosphere to form a "slow" neutron component. This chain of nuclear reactions is called the nucleonic component of the Cosmic rays and is composed of particles having relatively long lives.

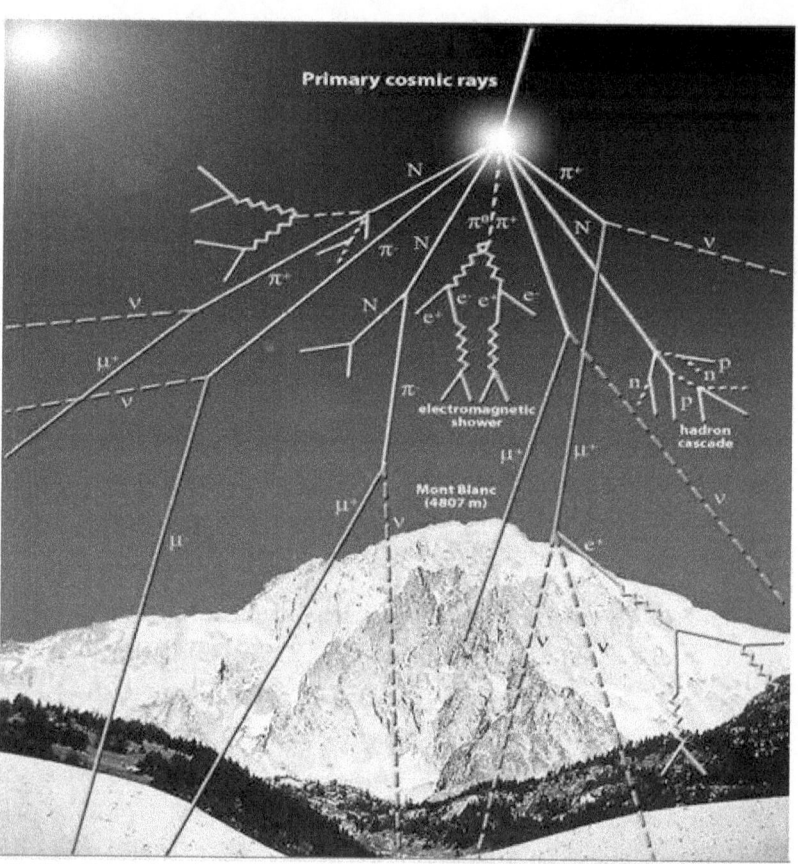

Fig 1.2 Penetration of Cosmic Rays Source :- physic.uibk.ac.at (from Quora.com)

1.5.2 Anomalous Cosmic Rays

Anomalous Cosmic Rays (ACR's) are cosmic rays with amazingly low energies. These types of Cosmic Rays are generated at the edge of the solar system ,in the heliosheath which is border zone between the heliosphere and the celestial medium. When electrically neutral atoms succeed in entering the heliosheath being unaffected by its magnetic field and eventually becomes ionized .They are accelerated into low energy cosmic rays by the solar wind's termination shock marking the inner edge of the heliosheath.Perhaps the high energy galactic Cosmic Rays hitting the shock front of the solar wind near the heliopause might be decelerated ,hence transforming into lower-energy anomalous cosmic rays. Low energy anomalous Cosmic rays were reported first time in 1973-74 during spacecraft observation made by space scientists(Garcia Munor et al 1973;Mc Donald et al 1974) .The energy flux of these particles ranged from 5 to 100 MeVN^{-1}.It has been observed that the energy of these low-energy anomalous cosmic rays increases with increasing distance from the Sun .These remarks established that low-energy anomalous cosmic rays have a nonsolar origin and is generated from the object located beyond the orbit of the Jupiter .Earth's magnetic field does not permit these low-energy cosmic ray particles to reach the earth's orbit. Low-energy charged particles are subtle indexes of solar activity and also of climate and weather. Most plentiful particles in the interplanetary space are protons and electrons ,having energy $\leq$ 5 MeV .Observations taken by some spacecrafts used for observing the low-energy anomalous cosmic rays like Pioneer 10 ,Pioneer 11 ,Voyager 1 & Voyager 2 have indicated that as the density of sun increases the density of these particles increases.

1.5.3 Solar Cosmic Rays

Solar Cosmic Rays are those particles associated with solar energetic particles (SEP's) or solar flares. SEP's are a type of Cosmic Rays which move away from the Sun due to heating of plasma, acceleration and many other forces. The solar flares often insert large quantities of energetic nuclei into space ,and the configuration varies from flare to flare.Cosmic rays are deflected by the magnetic field in celestial space ,they are also affected by the interplanetary magnetic field embedded in the solar wind and hence face a difficulty in reaching the inner solar system. As solar activity varies over the 11 year solar cycle the intensity of cosmic rays at earth also varies ,in anticorrelation with the sunspot number. The Sun is an occasional source of cosmic ray nuclei and electrons are accelerated by shock waves traveling through the corona ,and by magnetic energy released in solar flares.Such solar particle events are much more frequent during the active phase of the solar cycle.

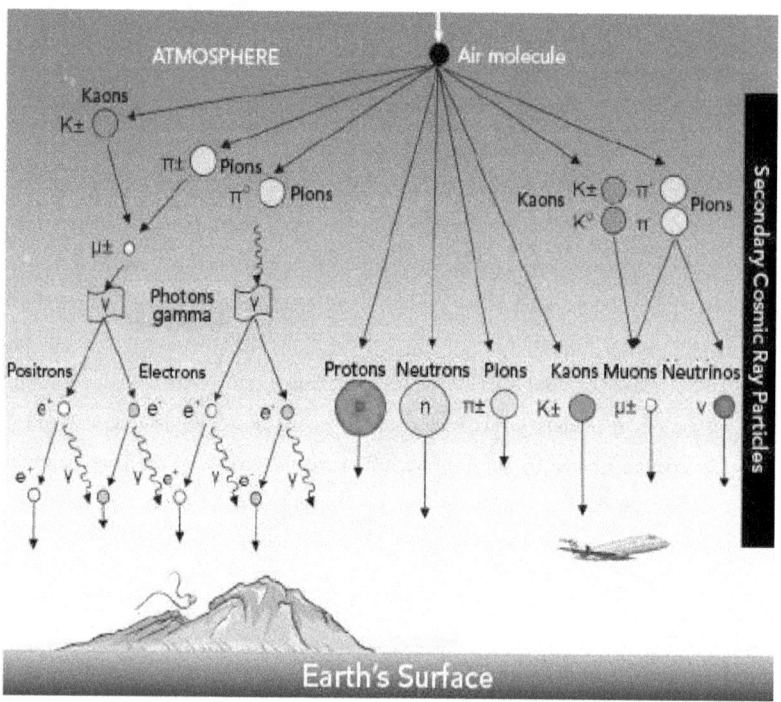

Fig 1.3 Classification of Secondary Cosmic Rays Source:-wikimapia

1.6 Cascade Development in the Atmosphere

Primary cosmic rays having energy $\geq 10^{13}$ eV, intruding at the top of the atmosphere basically are $\approx$ 92% protons, $\approx$ 6% alpha particles, and heavy nuclei and less than 1% of electrons and gamma rays. These particles reaching the earth's atmosphere, which undergo a series of interactions, leading to the generation of downward moving shower of particles. Atmospheric density with altitude essentially follows an exponential distribution and the mean free path for the Cosmic Ray protons and gamma ray photons are 80 gm. cm^{-2} and $\approx$ 38gm. cm^{-2} respectively. During the collision, as primary cosmic ray particles losses about 50% of their energy to a large number of secondary particles. The first interaction of the particles takes place rather deep in the atmosphere (typically at the altitude <30 km.). A gamma ray initiated pure electromagnetic cascade having different characteristics which compared as proton. In both the cases, interactions give rise to a laterally extended showers of relativistic electrons and positrons ($e^{\pm}$), called the "shower cascade". Gamma ray initiated showers would have predominately pair-production ($e^{\pm}$) process as the first interaction. The resulting ($e^{\pm}$) produces secondary gamma rays through bremsstrahlung interaction, these photons become the source particles for another generation of electron positron pairs, and so, eventually leading to the production of a fully developed showers of electron positron pairs and lower energy gamma rays, all traveling earthwards with ultra relativistic velocities. On the contrary, in the case of a cosmic ray proton, the first interaction is generally a nuclear interaction wherein a number of pions (°, +, -), kaons and other hadrons are produced, in addition to the e-m component, initiated by ° decay gamma rays. These showers are relatively richer in the muon component (decay product of charged pions), where as a purely electromagnetic shower (initiated by a gamma ray photons) is approximately exclusively made of electromagnetic components ($e^{\pm}$ gamma rays). Figure 1.3 shows the EAS development with primary particle energy $\sim 10^{12}$ eV or above and gives schematic representation of photons and nuclear cascade respectively while in their development phase in the earth's atmosphere. All the particles being of high energy move with velocities close to the velocity of light and hence the whole shower moves down the atmosphere as a disk of few meter thickness. The charged secondaries $^{\pm}$ and $K^{\pm}$ mesons having relativistic energy and mean life time ($\sim 10^{-8}$ sec) participating in the chain of nuclear interactions form the nuclear active component. On the other hand neutral pions (°) and kaons (K°) have very small life time ($\sim 10^{-15}$ sec and 10^{-10} sec respectively). The muons are highly penetrating particles because of their long life time (~2 x10^{-6} sec) and small interaction cross-section. The particles in a shower can be divided into three components

(a) the electromagnetic or simply the electron component, consisting of electron-positron and gamma rays

(b) the hadronic or nuclear active particle(NAP) component, consisting of all the strongly interacting particles hadrons and the muon component consisting of $\pm$arising from meson decays.

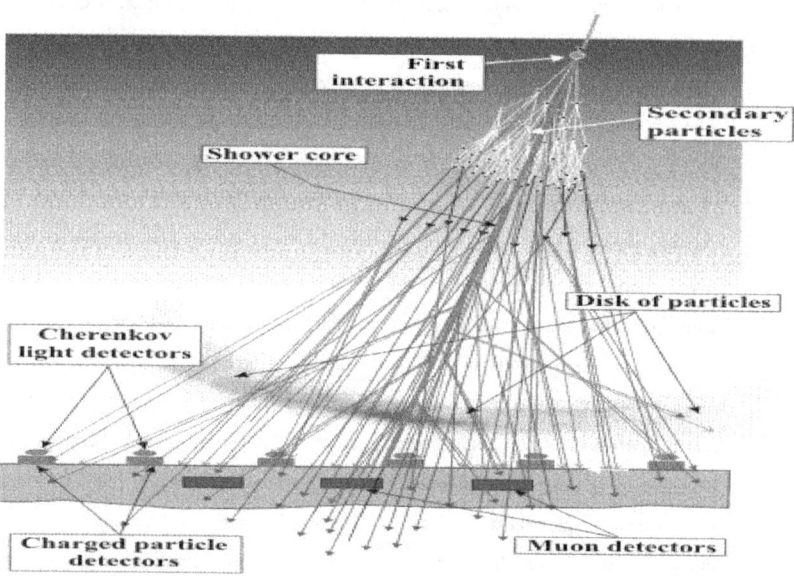

Fig 1.4 Role of Cosmic Rays in Atmosphere (Website: https://ikfia.ysn.ru/en/eas/)

1.6.1 ELECTROMAGNETIC COMPONENTS :

(a) Electromagnetic Cascade

When two hadrons collide each other neutral pions decomposes immediately into two high energy gamma photons almost in no time ($\sim 0.83 \times 10^{-16}$ Sec.)

$$\pi^0 \rightarrow \gamma + \gamma$$

The electromagnetic cascade is the result of a series of electromagnetic interactions of electrons and photons with the atmospheric nuclei. These increasing number of electrons and photons develop, as they travel down the atmosphere through a process called PHOTON-ELECTRON cascade.

Photons interact with matter through the following three processes.

(i) Photoelectric Effect.

(ii) Compton Effect,

(iii)   Pair Production.

Photoelectric effect is the process by which a photon losses its energy completely sending parts of its energy in liberating an electron from the atom and the rest is given to the liberated electron as kinetic energy. This process is important only at very low energies. The Compton effect is the process by which the photon interacts with a free electron or at energies high enough for the orbital electrons to be considered free. In this process photon transfers part of its energy and momentum to the free electrons immediately at rest and emerges with reduced energy, i.e. in Compton process

$$s\,\gamma \longrightarrow \gamma + e^-$$

Pair production is the process by which a photon interacts with the Coulomb field of the nucleus. In this interaction, the photon disappears and an electron-positron pair is created. The progression of secondary particles by gamma and proton initiated showers is shown in the figure 1.4.

This process can occur only at energies higher than the sum of the rest masses of electron and positron (~ 1.02 MeV), The positron annihilates producing more gamma photons.

$$e^+ + e^- \rightarrow \gamma + \gamma.$$

Again electrons interact with matter through the following process,

(a) Ionization and

(b) Bremsstrahlung.

The energy lost by an electron or a positron due to ionization in a medium produces more electrons. On the other hand, electron produce more gamma photons by Bremsstrahlung. In this process a high energy electron emits a photon when it interacts with the electromagnetic field of the nucleus in the medium. These processes are repeated producing electron-photon cascade and the overlap of the individual electron photon cascades of all the gamma rays produced above a particular level of observation constitute the electromagnetic component at that level. The electrons here include both the electrons and positrons. A fraction of the charged pions on the other hand decays into muons and neutrinos. Some of these muons decay by giving electrons, positrons while others reach the ground level. Muons are also produced as a result of decaying of W and Z bosons. A part of the charged pions collide with air nuclei giving rise to secondary showers. The process of electron-photon multiplication continues down the atmosphere till they reach the maximum after which absorption takes place.

1.6.2 Muon Component:

When pions & kaons decompose, they form muon component. Muons may also be formed by decay of charged particles through direct production process or from the decay of W and Z bosons. This component constitute about 10% of the total number of particles in the shower. Muons are nearly stable and have a small cross section for interaction, and they penetrate 12-14 km into the underground. Muons are called the "penetrating component" of the Cosmic Rays. They are relatively easy to detect because they are charged particles and reach the observation level directly from the point of the production. Only a few of them decay into electrons (or positrons) and neutrons during flight. High energy muons and also low energy muons at large distances from the core are produced at very high altitudes and hence carry important information about highest energy interactions in the cascade as well as the nature of the primary particle.They also carry genetic information from various stages of longitudinal development of Extensive Air showers. The average production height of the muons in a given shower increase with the energy of the muons. Therefore, by choosing the higher energy muons one can probe that region of the longitudinal development of the shower where interactions of energy higher than available by accelerators.Moreover, the total number of muons in a shower depends upon the energy per nucleon of the initiating primary rather than its total energy. Capdevielle et al propose that, primary energy can be estimated from the electron abundance by a simple formula.

$$E_0 = \left[a \ln\left(\frac{N_\mu}{N_e}\right) + b \right] N_e \text{------------------------(1.1)}$$

Where a = 0.404, b = 3.932 for $\frac{N_\mu}{N_e} < 0.666\%$

a = 1.534, b = 9.6, for $\frac{N_\mu}{N_e} \geq 0.666\%$

$$N_e < 25800$$

a = 1.88, b = 11.35, for $\frac{N_\mu}{N_e} \geq 0.666\%$

$$N_e > 25800$$

As the distance from the core increases, density of muons decreases. In fact, at sea level while 50% of all the particles in a shower are found inside a circle of radius ~ 70m, 50% of the muons are in a circle of radius ~ 300m. Thus the muons numbers, N_μ is a good index of primary energy, which can be written as,

$$Ep = 10^6 \left(\frac{N_\mu}{2 \times 10^4}\right)^{1.1} eV \text{--------------------(1.2)}$$

The energy spectrum of muons can be represented by a power law, as in the case of hadrons. Muons lose energy by ionization and by radioactive process, bremsstrahlung, direct production of $e^+ e^-$ pairs and photo nuclear interactions. It is flat at low energies because of increasing loses due to ionization and decay with decreasing muon energy, and becomes steeper at higher energies. The relation between the number of muons of energy $> E_\mu$ and the shower size can be expressed as,

$$N_\mu(> E_\mu) \propto N_e^{\alpha_\mu \mu(E_{\mu\mu})} \text{---------------------(1.3)}$$

The value of α_μ is 0.8 - 0.9 at $E_\mu \sim$ 200 GeV. This feature makes the muon component sensitive to the primary mass number. In the interaction mean free path, the number of muons at energy $> E_\mu$ in a shower given by,

$$N_\mu(A) \propto A \left(\frac{N_e}{A}\right)^{\alpha \alpha_\mu} \propto A^{1-\alpha_\mu} \text{-----------------------(1.4)}$$

Thus showers generated by heavy primaries have larger number of muons. The higher the energy of the muons, the more sensitive to primary mass number.

1.6.3 Hadron Component :

The hadron component which is also called the nuclear active particle (NAP) components includes nucleons, anti-nucleons, charged pions and kaons. This is the least abundant component of the shower constitutes about 1% of the EAS population, but carry substantial amount of energy of the primary particle and forms the backbone of the shower. Due to the confinement of the hadron components within 10 to 20 meters from the shower core, it is very difficult to study this components both at near the core as well as at further away from the core. At near the core, interference from the EM component is very large, while at far distance from the core. The density of hadrons is very small because of steep lateral distribution and small total number. A detailed study of hadron component associated with air showers of different sizes has been done by various experimental groups. It has been established that the hadron energy spectrum has a power law form with an exponent varying between - 1.2 & - 2.0 and its lateral distribution has an exponential form which flattens as the shower size increases. The agreement between various experiments is poor and is attributed to different sensitivities, energy and spatial resolutions and errors in energy estimation. The number of HE hadrons above any fixed energy in a shower varies almost linearly with shower size. The charged/neutral ratio $\left(\frac{C}{N}\right)$ and the time structure of hadrons as well as nucleon-anti nucleon pair production process in high energy interactions are studied well. For the production of kaons and nucleon-anti nucleon pairs, the ratio of $\left(\frac{C}{N}\right)$ can be expressed as,

$$\frac{C}{N} = \frac{nn_{\pi^\pm} + n_{k^\pm} + 0.5 n_{NN} + 0.5}{0.5 n_{NN} + n_{k_L^0 k_S^0} + 0.5} \quad \text{-----------------------------(1.5)}$$

The $\left(\frac{C}{N}\right)$ ratio for pions, kaons and nucleons is ∞, 1.5 (since k_s^0 are short lived and not many of them survive) and 1 respectively. By comparing the total number of hadrons and $\left(\frac{C}{N}\right)$ ratio with Monte Carlo simulations, the production cross- sections of different types of particles can be obtained.

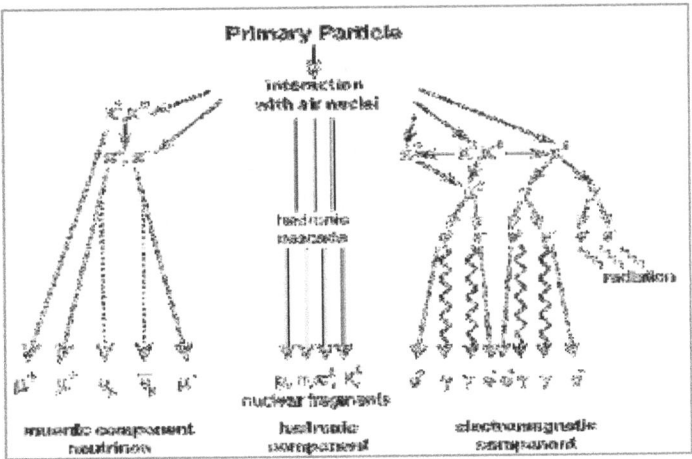

Fig 1.6 Schematic view of an extensive air shower (EAS), where KASCADE is measuring the hadron, muon, and electron components

1.7 Lateral and longitudinal development of charged particles :

In the beginning when high energy electron or photon arrives on the top of the atmosphere, Extensive Air Showers(EAS) occurs. The nucleonic origin of EAS was established when a more comprehensive study of its longitudinal and lateral development was made. The particles of EAS spread side ways to hundreds of meters due to the multiple coulomb scattering. The density of the shower particles at any level of observation is maximum at the core (the central axis) of the shower and it falls off rapidly with the distance from the shower axis. As the shower disseminates down the atmosphere, the hadron and the electromagnetic components increases in size, reach a maximum and then decrease while the muon component does not suffer significant depletion after reaching maximum because muons lose energy only by ionization and a small fraction of them are lost by decay. Extensive Air Showers are Complex Electromagnetic Cascade. An age parameter, s, is defined such that s = 1 at the shower maximum. Value of s < 1 corresponds to young showers, i.e. the shower which has not reached its maximum development, and s > 1 corresponds to old shower. The e..m. component is the most abundant component in air shower. The total number of charged particles in a shower is called the shower size (N) of that particular shower. On development, a shower also spreads laterally, i.e. perpendicular to the direction of the incident primary particles, due to the multiple Coulomb scattering. The photon and electron starts the longitudinal & lateral development of electron-photon cascade. Numerical results obtained by Kamata and Nishimura can be well approximated by a function, known as Nishimura - Kamata-Greisen (NKG) function, as suggested by Greisen.

$$\Delta^{(Ne,s,r)=} \frac{Ne}{2\pi r_0^2} \frac{\Gamma(4.5)}{\Gamma(s)\Gamma(4.5-2s)} \left(\frac{r}{r_0}\right)^{(s-2)} (1+\frac{r}{r_0})^{(s-4.5)} \ldots \quad (1.6)$$

Where $\Delta(N_e, s, r)$ gives the density of the particles per meter square at a distance r in a plane perpendicular to the shower axis in a shower with total number of particles (shower size) N_e, is the gamma function, s is the age parameter and $r_0 = \frac{E_s X_s}{\varepsilon_0 \rho}$ is the Moliere unit of length or "Scattering length" and "ρ" is the density of air at the observation level. Later, it has been shown that the NKG function does not give a good fit to the lateral distribution of electrons in the shower. For a shower initiated by a photon which is observed at a depth "t" (radiation length) from the point of origin, 'S' shower age is given by,

$$S = \frac{3t}{\left[t+2\ln\left(\frac{E_p}{\varepsilon_0}\right)\right]} \quad \text{-----------------------(1.7)}$$

Here, E_p is the primary energy, ε_0 is the critical energy for electrons in air which is equal to 84 MeV. It has been observed that 'S' is not constant across the lateral structure of the shower but decreases with the distance from the shower axis. The possible errors introduced in the estimation of shower size and age parameter due to use of NKG function in fitting the lateral distribution is discussed in detail by Capdevielle and Gawin.

On sampling the density of the shower at various points and then fitting the above lateral distribution function to these densities, N, s and co-ordinates of the shower core (r) i.e. (x_0, y_0) can be resolved. The total size N is a measure of the total energy of the primary particles. Fluctuation in development from shower to shower is large, even for shower of the same energy and primary mass. Thus the shower size N_e and the primary energy E_0 are only related in an average way and this relation depends on depth in the atmosphere. The relation for the vertical shower with 10^{14}eV < E < 10^{17} eV at 920 gmcm^{-2} (965m above s.l.) is

$$E_0 \sim 3.9 \times 10^6 GeV \left(\frac{N_e}{10^6}\right)^{0.9} \quad \text{-------------------(1.8)}$$

The lateral distribution function (LDF) for shower initiated by 100GeV gamma photons proposed by Hillas and Lapikens is in the form.

$$f_{HL} = C(s) \left[\frac{r}{0.25r_0}\right]^{a_1+a_2(s-1)} \left[1+\frac{r}{0.25r_0}\right]^{b_1+b_2(s-1)} \quad \text{----------------(1.9)}$$

Where $a_1 = -0.76$, $a_2 = 1.3$, $b_1 = -3.23$, $b_2 = 0$ and Moliere unit was reduced by a factor 0.25 of the usual 80m at sea level. Capdevielle et al, in the year of 1983 proposed a formula for radial electron density distribution.

$$\Delta_e(r) = C(s)\frac{N}{m^2 r_0^2}\left(\frac{r}{mr_0}\right)^{s-2}\left(\frac{r}{mr_0}+1\right)^{s-4}\left(1+d\frac{r}{mr_0}\right)^{2.7-s} \text{------------(1.10)}$$

Where d = 0.026 and m = 0.5

$$C(S) = 0.3265 \exp\left[-0.5\left(\frac{s-1.125}{0.499}\right)^2\right] \text{ for } s \leq 1.4$$

$$C(S) = 0.2854 \, S^2 - 1.385 \, S + 1.66 \text{ for } S > 1.4$$

The equation (1.14) can be written in the form,

$$\Delta_{cap}(r) = \frac{N}{m^2 r_0^2} f_{cap}\left(\frac{r}{mr_0}\right) \text{--------------------(1.11)}$$

This equation is a good approximation for lateral distribution function of electrons. Luorui et al reported a simple exponential function with the factor (r+1) instead of r in the denominator to avoid the divergence at r = 0, for the determination of density of shower particles as,

$$\rho = \frac{N\exp(-r/r_0)}{2\pi r_0 (r+1)} \text{--------------------(1.12)}$$

1.8 Cosmic Ray Modulation-

The flux of Cosmic rays incident on the Earth's upper atmosphere is modulated by two processes : the Solar Wind and the Earth's magnetic field. Solar Wind is expanding magnetized plasma generated by the Sun, which has the effect of decelerating the incoming particles and partially excluding some of the particles with energies below about 1 GeV. The amount of solar wind is not constant due to changes in solar activity over its regular eleven-year cycle. Hence the level of modulation varies in autocorrelation with solar activity. The intensity of Cosmic radiation depends on latitude, longitude and azimuth due to which the Earth's magnetic field diverts some of the Cosmic Rays. The Cosmic ray intensity at the equator is lower than at the poles as the geomagnetic cutoff value is greatest at the equator, due to which the cosmic ray particles tend to move in the direction of field lines and not across them. The longitude dependence arises from the fact that the geomagnetic dipole axis is not parallel to the Earth's rotation axis..Due to various dynamical processes occurring on the Sun and extending into interplanetary space, the modulation in Cosmic ray intensity takes place. The Cosmic Ray intensity variations can be divided into two categories long-term and short- term variations.

Long-term variation is classified as follows :-

- 11-year variations
- 22-year variations

Short-term variation is classified as follows :-

- 27 – Days variation
- Forbush Decrease
- Ground Level Enhancement

1.8.1 Long-term Variations-

Solar wind expands and flows continuously from the Sun into interplanetary medium, the magnetic field associated with it varies both in time and space according to solar conditions. Cosmic rays being charged particles are affected by the magnetic field variations on the scales of years. Two major variations in Cosmic ray intensity are those pertaining to the 11-year period of solar activity shown by the Sun Spot number and the 22-year period of solar magnetic polarity cycle.

11 & 22-Year Variations-

The Cosmic Ray flux is modulated by the 11-year solar cycle of Sun spot activity, reaching a maximum during the quiet period of solar cycle and minimum near the peak of solar activity i.e. the Cosmic ray intensity changes over the solar cycle in anti-correlation with Sun spot activity.Forbush has proved in 1954 that the mean Cosmic ray intensity is in anti-correlation with solar activity ,with an apparent period of 11 – years. Nagashima has proposed in 1977 about the existence of 22-year modulation as the result of the polarity reversal of the magnetic field of the Sun ,which occurred in the period of 1969-70 ,Howard et al. ,1974 declared that the boundary between the magnetic field of the Sun and the interstellar magnetic field considerably altered with the reversal of polarity of the dipole field on the Sun.The mechanism of 22-year variation is linked with the reversal of the dipole field of the Sun.In reference of the dipole field of the Sun ,there are two predominant concepts (i) Low Intensity States

High Intensity States

In High Intensity States, the galactic magnetic field and solar magnetic field are parallel to each other.This eases the entry of Cosmic ray particles inside the heliosphere.In Low Intensity States, the galactic magnetic field and solar magnetic field are oppositely pointed.In the 22-year cycle ,there are two different paths during the two different halves of the cycle which was also suggested by Jokipii and Thomas (1981). Jokipii and Thomas also suggested a drift model in which the hemispheric current sheet plays an important role the positive nuclei enter the heliosphere preferentially along the heliospheric current sheet when the northern solar magnetic field is positive. The neutron monitor data of 1957-1989 ,their results show sharp peak intensity in May 1981 similar to that of 1965 in contrast to flatter maximum observed in between 1972-77 etc. and are consistent with the predictions of the drift model having an equatorial hemispheric current sheet.The electron spectra for 1970-80 are sensitive to the change in the wavier of the current sheet ,whereas, spectra for the period 1980-90 is insensitive to that which is in contrast to the expectations of the drift model .

1.8.2 Short – Term Variations-

Since anisotropy of Galactic Cosmic Rays exists in the interplanetary medium ,it is cons idered as the daily variation of Cosmic Rays.The Short-term variations may be categorized as :-

(i) Symmetric Depressions

(ii) Asymmetric Depressions

These two types of depressions have come from whether the 'decrease' and 'recovery' parts in their time profile are 'symmetric' or ''asymmetric'. Symmetric Depressions have (a) V-shape and (b) Bowl or U-shape . Asymmetric Depressions are classified into

(i) Forbush decrease

(ii) Composite decrease

(iii) Wavy decrease

These types are dependent upon the time profile of the depression. They are also categorized as

(i) Forbush decrease

(ii) 27 days variation

(iii) Diurnal variation

(iv) Semidiurnal variation

(v) Higher harmonics

The first two types of variation are isotropic while the rest are anisotropic variations. Any solar parameter that influences cosmic ray intensity rotates with the rotation of Sun and reoccurs every 27 – days producing 27 – days variation.A short – term abrupt change in solar environment like solar flare affects the Cosmic Ray intensity for a short period of few days.These are called Forbush Decreases.A periodicity of 24 hours in Cosmic Ray intensity is called as Diurnal Variations.This is produced by the rotations of the Earth on its axis so that a detector of 24-hour recurrent variation sees any difference in flux in any particular direction.Harmonic Analysis technique is generally used to analyze diurnal variation.Higher order harmonic analysis called semi-diurnal variations and tri-diurnal variations .The existence of first three harmonics has been confirmed and investigated properly.The analysis of fourth harmonic is difficult to be performed as it is probable to have amplitude of maximum variations of around 0.01% this make the statistical analysis and observations practically impossible

(2) 27 – Day Variation-

Bukata et al. have enunciated in 1986 that twenty seven days variations is a quasi-permanent phenomenon and it is difficult to separate from Forbush decreases though generally their percentage of decrease is smaller than Forbush Decreases.Any observer on Earth will observe co- rotation stationary shock on Sun as a twenty-seven days recurring phenomenon.Decreased intensity of Cosmic Rays would be observed due to enhanced outward convection of particles along the line of force within a shock.Naskidas Shvilli et al. (1976) have found good agreement between theoretical predictions and practical observations based on calculations in three –dimensional interplanetary space and taking into account symmetry of solar wind boundary.It was suggested by Barouch and Burlaga in 1976 and confirmed by Duggal et al in 1983 that the drift associated with gradient in B_z may also be an important element in the predictions of these decreases.They have obtained a correlation between increase in B_z and reduction in Cosmic Ray intensity.Lucci et al. 1979 and Agrawal et al. 1980 have attributed the 27-days recurrent cosmic ray intensity decrease to the high speed solar wind streams coming out from coronal holes.

Forbush Decreases-

Forbush Decrease has been attributed as one of the major sporadic variations.It is abbreviated as (Fd).It is a transient and rapid decrease in the observed galactic cosmic ray intensity followed by a gradual recovery, typically lasting for several days.It has been acclaimed as one of the impressive Short-Term Variations.In Fd's event, the beginning of the decrease is quite abrupt and the minimum is reached within a day.Forbush Decreases occur when the Sun releases an exceptionally large burst of matter and magnetic disturbance also known as Magnetic Cloud.The disturbance sweeps away some of the Cosmic Ray energetic particles in its path and prevents many Cosmic ray particles from entering into the atmosphere.Earlier ,it was shown by J.A.Lockwood in 1971 that when these disturbances in the form of magnetic clouds passes over the Earth ,a Forbush decrease is seen on the particle detector.

The term "Forbush Decrease" was coined after the American physicist Scott E.Forbush in 1938,who studied Cosmic rays in the 1930's and 1940's.Till then,various investigators like (Webber et al 1981,Agrawal S.P. 1981,1983,Cane et al 1993)to name a few have instituted that Forbush decreases are produced by perturbation in the interplanetary condition and those perturbations emerged from shock waves,solar flares ,coronal mass ejections(CMEs) and flare generated high speed solar wind streams.many theories, depending on perturbations in the interplanetary conditions ,have been given by many investigators to explain Forbush decrease phenomenon but no one has emerged successful in explaining the details of Forbush Decrease (Fd).When the study of Cosmic rays started,based on the data obtained by ionization chambers, Forbush (1937) and Hess and Demmelmair (1937) observed decreases in the Cosmic ray count rate ,which usually lasted for a week.Simpson (1954) indicated that the origin of these decreases was possibly in the interplanetary medium.Afterwards,these were found to be pertaining to the geomagnetic storms and their driver ICMEs (Interplanetary Coronal Mass Ejections) magnetic clouds and corotating shock fronts.

There are two basic types of Forbush decreases (Cane,2000).They are:-

(a)*Non-recurrent decreases* caused by transient interplanetary events related to mass ejections from the Sun,are marked by a sudden onset,reach their maximum depression in about a day, and are characterized by a slow recovery.

(b)*Recurrent Decreases* consisting of a gradual onset and having a symmetric profile .and tend to be associated with co rotating high speed solar wind streams (Lockwood ,1971;Lucci et al,1979).

Ground Level Enhancements-

Ground Level Enhancements are abbreviated as GLE's.They are sudden increases in the Cosmic Ray intensity recorded by ground based detectors.GLEs are always associated with large solar flares but the acceleration mechanism producing particles of up to tens of GeV energy has not been understood till today.About 59 GLEs have been recorded since authentic records began in the 1940s and most current event was recorded on 16 July 2000.The increases in ground based measurements ranges from only a few per cent of background in polar monitors (with little or no geomagnetic cutoff) to 45 times for the 23 February 1956 event. The rate of GLEs would appear to be about one per year but there may be a slight clamping around solar maximum. For example, during 2 years centred on the last solar maximum 13 GLEs were recorded. Most solar flares associated with GLEs are located on the western section of the Sun where the IMF is well connected to the Earth. Because of its shape it is known as the "garden hose" field line. GLEs associated flares are located near to the foot point of the garden hose field line and are usually delayed in their arrival at earth and have more gradual increases to maximum intensity. It is very rare to observe GLEs associated flares to the east of the central meridian or Sun-Earth line. Although a large solar flare is invariably associated with GLE flare itself may not be causally related to the production of the high energy protons that produce the GLE response at Earth. The Solar energetic particle events are not rare and energetic protons are produced in common with CMEs and interplanetary shocks. These protons do not have sufficient energy to produce secondary particles that reach ground level but are clearly observed by spacecraft. Such CMEs and their associated shocks are most often produced without a solar flare. It is possible that there is a continuum to the acceleration process and that flares are energetic events. Alternatively, possibility of flare produces seed population of higher energy protons that are accelerated to GLE.

Anisotropic /Time Independent variations-

The spatial anisotropy of the galactic cosmic radiation in the interplanetary medium is observed as the daily variation of cosmic ray intensity by the ground based detectors. The detector situated on earth scans the entire celestial sky during a period of 24 hours since the Earth completes one revolution around its own axis once in 24 hours (a solar day). Many scientists including Brunberg and Dattner (1954), using the ground based detector system, excess of particles arriving from the asymptotic direction . The amplitudes and phases of the harmonic components are generally obtained by Fourier analysis by using the hourly counts . A large difference in amplitude or time of maximum between different stations shows that either the anisotropy is changing within 24 hours or the universal time effects are predominant. At Earths, the neutron monitors respond to an energy range from 1 .5

GeV to a few hundred of Ge the existence of I, II and III harmonics of daily variations of cosmic rays have already been confirmed by various investigators. Their characteristics have been investigated by various workers both on theoretical as well as on experimental basis. In the neighbourhood of Sun, the particles. corrotates with the Sun, higher cosmic ray intensity is observed from a direction opposite to Earth's orbital motion. Based on the concept of continuous emission of the solar wind and stretching with it the smooth IMF Parker (1961). Ahluwalia and Dessler (1962), proposed a model which could partially explain the average behaviour of diurnal variation.) According to this model, the cosmic ray particles, drift due to the electric field E (E=V x B). The model predicts that (i) The source of the diurnal variation which is in the ecliptic plane is energy independent $(\beta = 0)$.(ii) All particles, whose rigidity is below an upper cut off rigidity ($R_{max} \approx 100\,GV$), undergo complete correlation (iii) The direction of anisotropy is perpendicular to the IMF direction. The amplitude of the anisotropy is proportional to the solar wind velocity.

Diurnal Variations of Cosmic Rays-
Solar diurnal anisotropy is caused by solar moderation of the galactic Cosmic rays in the heliosphere (Parker, 1964; Forman and Gleeson, 1975). Cosmic rays enter the heliosphere and revolve along interplanetary magnetic field (IMF) line. Magnetic flux irregularities scatter cosmic rays from their gyro-orbits, causing diffusion. The solar wind also convects cosmic rays outward, while particles travelling along a regular portion of magnetic field lines undergo magnetic curvature and drifts Isenberg and Jokipii (1979). The solar diurnal of the Cosmic-ray intensity was studied and interpreted initially on the basis of an outward radial convection and inward diffusion along the interplanetary magnetic field. The balance between convection and diffusion generates an energy-independent anisotropic flow of cosmic ray particles from the 18-hour co-rotational direction is produced. Bieber and Chen (1991) investigated the long-term behaviour of the cosmic ray diurnal variation, as observed by a series of neutron monitors over the period 1936 to 1988.They have found that the amplitude of the diurnal variation shows an 11-year periodicity, but the phase of this variation has 22-year cycle. According to the standard diffusion and convection theory the average characteristics of the diurnal anisotropy have notably changed during 1971-74 and phase shifted to earlier hours since

1971 (Rao et al, 1972; Forman and Gleeson, 1975, Agrawal and Singh, 1975; Kumar, 1978).The amplitude and phase of the diurnal anisotropy also changes with a period of one or two solar cycles (Rao, 1972, Singh and Badruddin, 2007).

Semi-Diurnal variation of Cosmic Rays-

The presence of a semi- diurnal component was confirmed first time by Ables et al (1965) by means of numerical filter techniques to improve the direction of small signals in the presence of statistical fluctuations or noise. By making use of the hourly values of neutron and meson detectors located on earth many explorers have studied the characteristic properties of semi-diurnal variations. The existence of semi- diurnal variation is due to the particle density gradient in the plane perpendicular to solar equatorial plane (Lietti and Qenby 1968). In current research, the significant correlation between semi and tridiurnal amplitudes has been observed on monthly/ yearly average basis while amplitudes of semi- diurnal component do not show significant correlation with the amplitudes of diurnal component of cosmic ray intensity variation. Agrawal et al. (1981) ; Ahluwalia et al. (1991) ; and EI- Borie et al. (1995) outlined a systematic solar cycle and solar wind dependence of semi- diurnal components during a long period of 1953-1987. Their results showed that there are three enhancement of declining phases of solar activity cycles (19,20 and 21), which are associated with high values of solar wind velocity. Their results also confirm the dependence of solar daily variation on solar wind velocity.

Tri and Quart Variations of Cosmic Ray Intensity -

Nagashima et al. (1977) have delineated the importance of quart diurnal variation of Cosmic rays. By making use of the data of high counting rate multi-direction telescope of the Norikura without describing the significant characteristics. Tri- diurnal (third harmonic) and Quart diurnal (Fourth harmonic) variation of Cosmic ray intensity are generally known as higher harmonics. The extraterrestrial nature of tri- diurnal variation of galactic Cosmic radiation has been established only (observed amplitude $\geq$ 0.02%). Characteristics properties of tri-diurnal variation have been delineated by a large quantity of examiners using the data of neutron and meson monitors (Agrawal 1981). The average tri-diurnal amplitude enhancement observed by all the detectors during 1973-75 provide sufficient signal to noise ratio to obtain its variation characteristics. Agrawal et al (1981) have performed the analysis to derive the characteristics properties of quart diurnal variation of Cosmic rays, but did not found any meaningful results due to its low amplitudes, which are insignificant for any statistical analysis and conclusions. At the most the characteristic properties of the tri-diurnal variation have been reported as not very different from that of semi-diurnal variation supporting their common origin as required in some of the theoretical models Nagashima et al (1977). Agrawal (1981) have reported significant correlation between the theoretical and semi-diurnal amplitude for both muons and neutrons. Recently, Shrivastava and Singh 2011 reported the characteristics properties of Tri-diurnal variation of cosmic ray intensity for 23 solar cycle. Firstly the fourth harmonic (quart diurnal variation) of cosmic ray intensity was delineated by Nagashima et al. (1977) by making use of high counting rate data of multi-directional telescope at Mt. Norikura during 1963-75.However, its presence with significant amplitudes has not been observed by others, who have analyzed the neutron and meson monitor data. Agrawal et al (1981) have studied the characteristics of existence of quart diurnal variation of Cosmic ray intensity.

CHAPTER 2
LITERATURE REVIEW

2.1 Introduction

Cosmic rays are high energy charged particles, originating in outer space, that travel at nearly the speed of light and strike the Earth from all directions. Cosmic rays are mainly galactic cosmic rays originating from sources outside the solar system, distributed throughout our Milky Way galaxy. They include high energy electrons, positrons, and other subatomic particles. They also include other classes of energetic particles in space, including nuclei and electrons accelerated due to energetic events on the Sun (called solar energetic particles), and particles accelerated in interplanetary space. Cosmic rays are mainly composed of nuclei of atoms, ranging from the lightest to the heaviest elements in the periodic table. They may produce showers of secondary particles that penetrate into the earth's atmosphere and it reaches on the surface of the earth .Secondary Cosmic Rays are composed primarily of high-energy protons and atomic nuclei . A significant fraction of primary cosmic rays originate from the supernovae of massive stars as well as from active galactic nuclei. Cosmic rays were discovered in 1912 by Victor Hess et al., when he found that an electroscope discharged more rapidly as he ascended in a balloon. He told that the reason for it is a source of radiation entering the atmosphere from above, and in 1936 was awarded the Nobel prize for his discovery.

Theodore Wulf et a., 1909 developed electrometer to measure the rate of ion production at the top of the Eiffel Tower. In previous observation variations at the rate of ionization over the sea at a depth of 3m from the surface and concluded that ionization must decrease under the sea water. In 1912, Victor Hess found that the ionization rate increased approximately fourfold over the rate at ground level. Hess ruled out the Sun as the radiation's source by making a balloon ascent during a near-total eclipse. With the moon blocking much of the Sun's visible radiation, Hess still measured rising radiation at rising altitudes. He concluded "The results of my observation are best explained by the assumption that a radiation of very great penetrating power enters our atmosphere from above."In the 1920s the term "cosmic rays" was coined by Robert Millikan who made measurements of ionization due to cosmic rays from deep under water to high altitudes and around the globe. In 1927, J. Clay found evidence, later confirmed in many experiments, of a variation of cosmic ray intensity with latitude, which indicated that the primary cosmic rays are deflected by the geomagnetic field and must therefore be charged particles, not photons. In 1929, Bothe and Kolhörster discovered charged cosmic-ray particles that could penetrate 4.1 cm of gold. Since then, a wide variety of potential sources for cosmic rays began to surface, including supernovae, active galactic nuclei, quasars, and gamma-ray bursts. In common scientific usage high-energy particles with intrinsic mass are known as "cosmic" rays .Cosmic rays are of two types ,Primary Cosmic Rays and Secondary Cosmic Rays. Primary cosmic rays originate outside the Earth's atmosphere and sometimes even in the Milky Way. About 99% of the Primary Cosmic Rays are the nuclei (stripped of their electron shells) of well-known atoms, and about 1% are solitary electrons (similar to beta particles). Of the nuclei, about 90% are simple protons, i. e. hydrogen nuclei; 9% are alpha particles, and 1% are the nuclei of heavier elements, called HZE ions. A very small fraction are stable particles of antimatter, such as positrons or antiprotons. When they interact with Earth's atmosphere, they are converted to secondary particles. Secondary cosmic rays, caused by a decay of primary cosmic rays as they impact an atmosphere, include neutrons, pions, positrons, and muons. Of these four, the latter three were first detected in cosmic rays.The mass ratio of helium to hydrogen nuclei, 28%, is similar to the primordial elemental abundance ratio of these elements, 24%.When cosmic rays enter the Earth's atmosphere ,they collide with atoms and molecules, mainly oxygen and nitrogen,producing secondary cosmic rays . The interaction produces a cascade of lighter particles, a so- called air shower secondary radiation that rains down, including x-rays, muons, protons, alpha particles, pions, electrons, and neutrons. All of the produced particles stay within about one degree of the primary particle's path.Typical particles produced in such collisions are neutrons and charged mesons such as positive or negative pions and kaons. Some of these subsequently decay into muons, which are able to reach the surface of the Earth, and even penetrate for some distance into shallow mines.The flux of incoming cosmic rays at the upper atmosphere is dependent on the solar wind,the Earth's magnetic field, and the energy of the cosmic rays. At distances of ~94 AU from the Sun, the solar wind undergoes a transition, called the termination shock, from supersonic to subsonic speeds. The region between the termination shock and the heliopause acts as a barrier to cosmic rays, decreasing the flux at lower energies ($\leq$ 1 GeV) by about 90%. However, the strength of the solar wind is not constant, and hence it has been observed that cosmic ray flux is correlated with solar activity. In addition, the Earth's magnetic field acts to deflect cosmic rays from its surface, giving rise to the observation that the flux is apparently dependent on latitude, longitude, and azimuth angle. The magnetic field lines deflect the cosmic rays towards the poles, giving rise to the aurorae.

Cosmic rays causes damage on microelectronics and life outside the protection of an atmosphere and magnetic field, and scientifically ,because the energies of the most energetic ultra high energy cosmic rays have been observed to approach 3 x 10^{20} eV, about 40 million times the energy of particles accelerated by Large Hadron Collider ,cosmic rays attract great interest . Because cosmic rays are electrically charged they are deflected by magnetic fields, and their directions are random hence it is impossible to tell from where they originated. The number of particles reaching the Earth's surface is related to the energy of the cosmic ray that struck the upper atmosphere. Cosmic rays having energies beyond 10^{14} eV are studied with large "air shower" arrays of detectors distributed over many square kilometers that sample the particles produced. The frequency of air showers ranges from about 100 per m^2 per year for energies >10^{15} eV to only about 1 per km^2 per century for energies beyond 10^{20} eV. Most secondary cosmic rays reaching the Earth's surface are muons, with an average intensity of about 100 per m^2 per second. Although thousands of cosmic rays pass through our bodies every minute, the resulting radiation levels are relatively low, corresponding, at sea level, to only a few percent of the natural background radiation. However, the greater intensity of cosmic rays in outer space is a potential radiation hazard for astronauts, especially when the Sun is active, the interplanetary space suddenly gets filled with solar energetic particles. Cosmic rays are also a hazard to electronic instrumentation in space; impacts of heavily-ionizing cosmic ray nuclei causes computer memory bits to "flip" or small microcircuits to fail. Just as cosmic rays are deflected by the magnetic fields in interstellar space, they are also affected by the interplanetary magnetic field embedded in the solar wind (the plasma of ions and electrons blowing from the solar corona at about 400 km/sec), and therefore have difficulty reaching the inner solar system. Spacecraft venturing out towards the boundary of the solar system have found that the intensity of galactic comic rays increases with distance from the Sun. As solar activity varies over the 11 year solar cycle the intensity of cosmic rays at Earth also varies, in anti-correlation with the sunspot number. The Sun is also a sporadic source of cosmic ray nuclei and electrons that are accelerated by shock waves traveling through the corona, and by magnetic energy released in solar flares. During such occurrences the intensity of energetic particles in space can increase by a factor of 10^2 to 10^6 for hours to days. Such solar particle events are much more frequent during the active phase of the solar cycle. The maximum energy reached in solar particle events is typically 10 to 100 MeV, occasionally reaching 1 GeV (roughly once a year) to 10 GeV (roughly once a decade). Solar energetic particles can be used to measure the elemental and isotopic composition of the Sun, thereby complementing spectroscopic studies of solar material. A third component of cosmic rays, comprised of only those elements that are difficult to ionize, including He, N, O, Ne, and Ar, was given the name "anomalous cosmic rays" because of its unusual composition. Anomalous cosmic rays originate from electrically-neutral interstellar particles that have entered the solar system unaffected by the magnetic field of the solar wind, been ionized, and then accelerated at the shock wave formed when the solar wind slows as a result of plowing into the interstellar gas, presently thought to occur somewhere between 75 and 100 AU from the Sun (one AU is the distance from the Sun to the Earth).

Cosmic rays can cause mutations and changes in human genes .Cosmic rays can change our DNA make-up by hitting individual cells ,and may have human evolution implications .Astronauts are hitted by the cosmic rays in space when high energy electrons are stopped suddenly inside the astronaut's spaceship producing x-rays which can be harmful to humans if they are exposed to them for any length of time .

2.2 REVIEW OF LITERATURE

Solar cycle takes place in every 11 years .In Solar Cycle 22 (peaking in 1990), the dip in amplitude of 1.28% in the low altitude cloud cover has been seen.The huge thermal time constant of the outer part of the Sun limits the variability in its surface temperature , and hence its total power output ,which is dominated by visible and infrared emissions from the solar surface (the photosphere) (Lockwood et al. ,2004).The Galactic Cosmic Rays (GCR's) are very energetic .The expansion of the shielding solar magnetic field into interplanetary space results in the Sun modulating the number of GCR's reaching earth (Potgieter et al., 2008).Air ions generated by GCR's enable earth's global electric circuit (thunderstorm) and they also modulate the formation of low altitude clouds (Svensmark and Friis-Christensen et al. ,1997). The Sun also emits a continuous stream of low-energy charged particles called the solar wind (Marsch et al.,2006).A small fraction of the solar wind energy incident on earth is extracted by the geomagnetic field and deposited in the thermosphere at high latitudes (Cowley et al. ,1991 ; Thayer and Semeter et al. ,2004) .A decrease in Cosmic Ray(CR) intensity causes a decrease in Low Cloud Cover (LCC) .Such a causal connection has vast importance ,as it could be the main cause of the presently observed global warming.

Both electromagnetic and charged particle emissions from the Sun varies over the decadal-scale solar magnetic activity cycle ,like GCR fluxes .But any effects on climate are much more significant for any variations over longer timescales .Only empirical analogue forecasts of future solar output changes can be made as there are no predictive models of the solar dynamo (Weiss and Thompson et al. ,2009).There is an 8% chance that Sun will return to Maunder Minimum conditions within 50 years .The recent evolution of solar cycle 24 indicates that the Sun may be following such a trajectory .Predictions have shown that continued solar decline will do little alleviate anthropogenically driven global warming .Understanding spectral irradiance variability and differentiating "top-down" and "bottom-up" solar forcings will be needed as these will have very different effects on the spatial patterns of the responses and will behave differently in combination with other changes such as sea ice loss .

The climate reflects a complex and dynamical interaction between the prevailing states of the atmosphere ,oceans, land masses,ice sheets and biosphere ,in response to solar insolation .It is important to identify the primary forcing agents as they provide the fundamental reason for the climate change ,whereas the feedbacks determine by how much .Internal forcing agents (those arising within earth's climate system) include volcanoes ,anthropogenic green house gases and ,on very long time scales (tens of millions of years),plate tectonics .There are only two established external forcing agents : orbitally-modulated solar insolation and variations of solar irradiance. A linkage between orbit and climate is provided by the Milankovitch model, which states that retreats of the northern ice sheets are driven by peaks in northern hemisphere solar insolation.High precision paleoclimatic data have revealed serious discrepancies with the Milankovitch model that either fundamentally challenges its validity or, at the very least, call for a significant extension .These light radioisotope archives record the galactic cosmic ray (GCR) flux in Earth's atmosphere,which is modulated by the interplanetary magnetic field and its inhomogenities carried by the solar wind.Long term solar magnetic variability has been assumed to be a proxy for irradiance variability . This assumption lacks a physical basis, and more recent estimates suggest that long-term irradiance changes are probably negligible .

Earth's climate varies substantially on centennial and millennial time scales .Some of these variations are due to "unforced" internal oscillations of the climate system, involving components with suitably long response times ,such as the ice sheets.It is hard to explain away all these climate variations as internal oscillations-escpecially in cases involving large climate shifts (e.g. persistent suborbital variations of sea-level by 10-20 m have occurred during both glacial and interglacial climates , or where synchronous climate changes are observed in widely separated geographical locations ,without any clear path for their teleconnection .At present there is no established natural forcing agent on these time scales;they are either too short (solar irradiance,volcanoes) or too long (solar insolation,plate tectonics).Recently,numerous studies of centennial and millennial scale climate change have reported association with GCR variations .These are frequently considered as a proxy for changes of solar irradiance (or a spectral component such as solar ultra violet) . The ambiguous interpretation as either a solar-climate or a GCR-climate forcing mechanism can in principle be resolved by examining climate change on different time scales since, unlike solar irradiance, the GCR flux is also affected over longer times by geomagnetic and galactic variations, and over shorter times by solar magnetic disturbances. The complexity of the climate system implies that it is not easy to explain correlations of solar variability and climate in mechanistic models. The key challenge is to establish a physical mechanism that could link solar or cosmic ray variability with the climate. Since the energy input of GCRs to the atmosphere is negligible—about 10^{-9} of the solar irradiance, or roughly the same as starlight—a substantial amplification mechanism would be required.Since low clouds are known to exert a strong net radiative cooling effect on Earth, this would provide the necessary amplification mechanism—and also the sign of the effect: increased GCRs should be associated with cooler temperatures.Studying past climate variability, before any anthropogenic influence, is a good way to understand natural contributions to present and future climate change. If there is good evidence for a significant influence of solar/GCR variability on past climate then it is important to understand the physical mechanism.

Cosmic ray forcing of the climate could in principle operate on all time scales from days to hundreds of millions of years, reflecting the characteristic time scales for changes in the Sun's magnetic activity, Earth's magnetic field, and the galactic environment of the solar system. Moreover the climate forcing would act simultaneously, and with the same sign, across the globe. This would both allow a large climatic response from a relatively small forcing and also give rise to simultaneous regional climate responses without any clear teleconnection path. The most persuasive palaeoclimatic evidence for solar/GCR forcing involves sub-orbital (centennial and millennial) climate variability over the Holocene, for which there is no established forcing agent at present. Increased GCR flux appears to be associated with a cooler climate, a southernly shift of the ITCZ (Inter Tropical Convergence Zone) and a weakening of the monsoon; and decreased GCR flux is associated with a warmer climate, a northernly shift of the ITCZ and a strengthening of the monsoon (increased rainfall). The influence on the ITCZ may imply significant changes of upper tropospheric water vapour in the tropics and sub-tropics, potentially affecting both long-wave absorption and the availability of water vapour for cirrus clouds.

CHAPTER 3
EFFECT OF SOLAR ACTIVITY ON COSMIC RAY MODULATION

3.1 Introduction

The cosmic ray modulation is the changes in cosmic ray intensities with the 11-year periodical changes in solar activity from maximum to minimum .Marsh & Svensmark (2000) explained that the modulation of galactic cosmic rays in the heliosphere follows a 11 year modulation cycle and it is anticorrelated with solar activity with maximum cosmic-ray intensity values being obtained for solar-minimum conditions .Apart from the 11 year modulation cycle, 22 year cycle exhibited maximum intensity.The galactic cosmic ray modulation in the heliosphere occurs mainly due to the following processes: 1) particle diffusion perpendicular to the **Heliospheric Magnetic Field (HMF)**; 2) Due to gradient , curvature ,**Heliospheric Current Sheet(HCS)** and drift effects; 3) Convection and adiabatic energy losses in the expanding solar wind. These processes depends on the geometrical structure, polarity, strength, and turbulence level of the **Heliospheric magnetic field** and the **solar-wind speed** .Jacobs and Cohen in 1997 studied that the magnetic axis of the Sun is tilted with respect to its rotational axis, due to this tilt occurring solar cycle and leading variations in the heliospheric characteristics. Since the heliospheric characteristics vary in time depending on the phase of the **solar magnetic cycle** and increased solar activity level, so that resultant modulation of GCRs will also change with time. Heliospheric parameters provides the necessary backbone to understand space environment and for monitoring the forecasting of Space weather. Heliospheric study explains the negative correlation between individual Solar parameters and Cosmic Ray Intensity . The cosmic ray modulation from the perspective of heliospheric parameters will prove crucial.

The variation of CRI modulation with heliospheric parameters like Sunspot Number ,**Interplanetary Magnetic Field,Solar Wind Velocity** and **F10.7 (Solar Radio Flux)** has been shown by tracing Multiple Regression technique which provides information about plasmasphere and atmospheric earth. The strength of anti-correlation is different for different phases of solar cycle and varies with time and space. Variation in sunspot number is the indicator of solar activity which inturn depends on solar dynamo process.The modulation of galactic cosmic rays due to diffusion,convection and **adiabatic deceleration** leads to characteristic differences between adjacent solar cycles having different polarity of the solar and large scale **interplanetary magnetic fields**. The polarity of the **solar magnetic field** reverses sign once every 11-year near the time of maximum solar activity. Thus successive activity maxima are characterized by different solar field polarity.For a better understanding of odd–even solar cycle's differences, the influence of curvature of **interplanetary magnetic field** on the transport of cosmic ray should also be considered. The modulation of cosmic rays also depends upon the charge and sign of the **solar magnetic field** polarity during the 11 and 22 year modulation cycles. A large scale change in the magnitude of **solar magnetic field** has a profound impact on the drift experienced by the cosmic rays even they enter into the magnetosphere. It also affects the tilt experienced by the **heliospheric current sheet**.

3.2 Factors Affecting Galactic Cosmic Ray Modulation
1. Particle Diffusion Perpendicular to Heliospheric Magnetic Field

A collective phenomenon, which is a trait for larger particle communities, is termed as Diffusion. When encompassed or stochastic force on particles is applied by molecules of suspending medium ,diffusion is caused by its warm movement .The rate of diffusion increases with temperature and decreases with particle size and the viscosity of the medium A large number of high-energy phenomena in astrophysical environments, such as the transition of cosmic rays, interplanetary transport of solar energetic particles, gaining of momentum of charged particles by shock waves, and propagation of galactic cosmic rays in the interstellar medium are governed by the diffusion and drift of energetic charged particles in turbulent magnetic fields. The diffusion of energetic charged particles in a turbulent magnetic field is often written in terms of two coefficients, $k_\parallel$ and $k_\perp$, which respectively describe the diffusion parallel and perpendicular to the magnetic field. It was explained 30 years ago that particle diffusion normal to heliospheric magnetic field can be explained by the intertwining of the magnetic lines of force due to turbulence in the magnetic field. Particles follow individual field lines uncomposedly intertwined on a variety of length scales.

2. Gradient , Curvature ,Heliospheric Current Sheet(HCS) and Drift Effects

The gradient of H at a point is a plane vector pointing in the direction of the steepest slope or grade at that point. The steepness of the slope at that point is given by the magnitude of the gradient vector. Intuitively, the curvature is the amount by which a curve deviates from being a straight line, or a surface deviates from being a plane. For curves, the canonical example is that of a circle, which has a curvature equal to the reciprocal of its radius. The **heliospheric current sheet** is the surface in the Solar System where the polarity of the Sun's magnetic field changes from north to south. ... The shape of the **current sheet** results from the influence of the Sun's rotating magnetic field on the plasma in the interplanetary medium (solar wind). The gradient of a heliospheric magnetic field occurs due to the polar coronal hole being in equilibrium with the solar magnetic axis, and the magnetic field expanding nonradially to yield a uniform field farther away from the Sun. The footpoints of the magnetic field lines anchored in the photosphere experience differential rotation. The magnetic axis of the Sun and assumed that it rotate rigidly at the equatorial rate, differential rotation causes a footpoint to move in heliomagnetic latitude and longitude, thus experiencing different degrees of nonradial expansion.The end result is a field line that moves in heliographic latitude, and the simple concept of "field lines on cones" of the Parker field breaks down.In the transport of the various species of galactic cosmic rays (CRs) ,drifts due to gradients and curvatures in the heliospheric magnetic field (HMF), and along the wavy current sheet, are important processes that have a tremendous effect on cosmic-ray modulation, accounting for the observed 22-year cycle in cosmic-ray intensities. Drift effects, however, have been shown both theoretically and by means of numerical simulations to be reduced in the presence of turbulence Given the importance of drift in any study of cosmic-ray modulation, this reduction needs to be carefully modelled throughout the heliosphere.

3. Convection And Adiabatic Energy Losses In The Expanding Solar Wind

The impact of the solar wind onto the magnetosphere which generates an electric field within the inner magnetosphere (r < 10 a; with a the Earth's radius) - the convection field- . Its general direction is from dawn to dusk. The co-rotating thermal plasma within the inner magnetosphere drifts orthogonal to that field and to the geomagnetic field B_o. The generation process is not yet completely understood. One possibility is viscous interaction between solar wind and the boundary layer of the magnetosphere (magnetopause). Another process may be magnetic reconnection. Finally, a hydromagnetic dynamo process in the polar regions of the inner magnetosphere may be possible. Direct measurements via satellites have given a fairly good picture of the structure of that field. A number of models of that field exists.

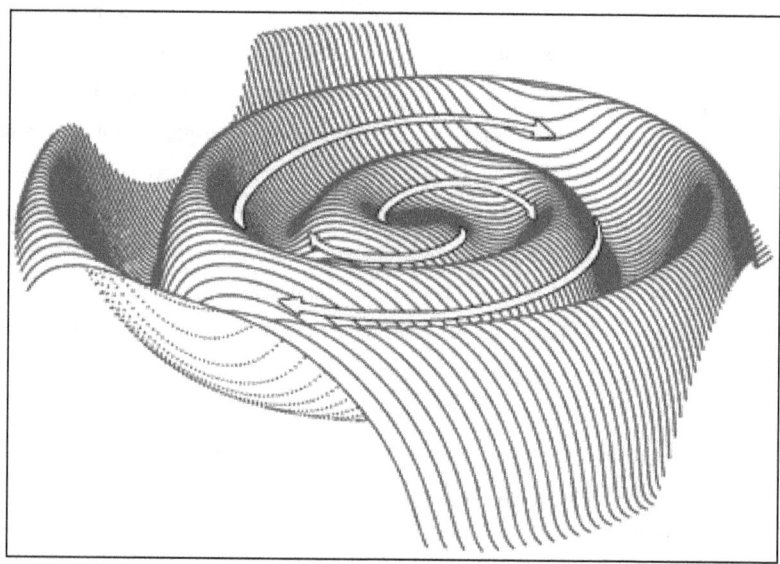

Fig 3.1 Heliospheric Current Sheet

3.3 Data Analysis

The GCR modulation study during Solar cycle 24 has been done by using pressure-corrected monthly averaged GCR intensity values of Oulu neutron monitor station (cutoff rigidity 0.81 GV; 65.05° N; 25.47° E) (http://cosmicrays.oulu.fi). In the present study, following parameters have been analyzed (i) Sunspot numbers (SSN), (ii) Interplanetary magnetic field (IMF) (iii) Solar wind velocity (iv)The F10.7-cm solar radio flux. The monthly data of sunspot numbers, interplanetary magnetic field, solar wind velocity and F10.7 –cm (2800 MHz) solar radio flux were taken from the Data Facility website of Omniweb Goddard Space Flight Centre (www.omniweb.gsfc.nasa.gov). The cosmic ray intensity data has been taken from Oulu Neutron Monitor Station which is a part of ReSoLVE National Centre of Excellence which is affiliated by Space Climate Research Unit and Sodankylä Geophysical Observatory (University of Oulu, Finland). Neutron Monitor Station of **Oulu** has 9NM64 neutron monitor with muon telescope in **Concordia** Station (Antarctica).Double mini neutron monitor collects real time cosmic ray intensity data .Dissemination of data have been done from the Omniweb Browser which gathers data from the IMP1 spacecraft held as part of the OMNI data set. The OMNI data set is an hourly-resolution data set with near-Earth solar wind magnetic field and plasma data, solar and geomagnetic activity indices, and energetic particle flux data since 1963. The data set is created by interspersing, after cross-normalizing, each data set of several spacecraft. The OMNIWeb interface provides access to this data set with graphical browse and sub-setting capability, to ftp-accessible annual ASCII OMNI files, and provides higher resolution data sets contributing to OMNI. Solar data have been taken since the early 1970's with the help of Pioneer 10 and 11 and IMP 8 spacecrafts and covers Heliospheric, Magnetospheric, Ionospheric and Upper-Atmospheric data from all NASA and some non-NASA space physics missions of the subsequent years. Some data services yield simple file retrieval, while others provide data sub-setting, graphical display, etc. Many data sets are available through the Coordinated Data Analysis Web (CDAWeb) service. These are predominantly magnetospheric data and nearby solar wind data of the ISTP era at time resolutions of approximately a minute. The CDAWeb service provides graphical browsing, data subsetting, screen listings, files creations and downloads (ASCII or CDF) files.

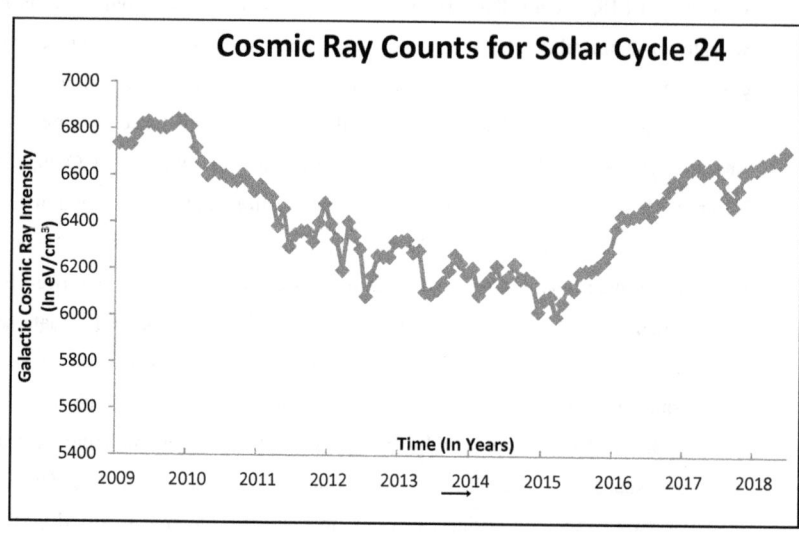

Fig 3.2 Yearly Variation of Monthly averaged Galactic Cosmic Ray Counts during Solar cycle 24 (December 2008 – June 2018)

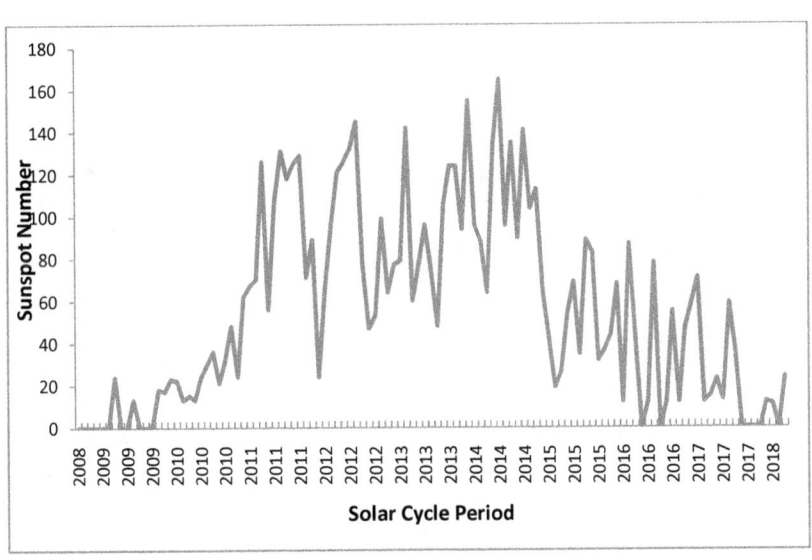

Fig 3.3 Variation between Monthly averaged Sunspot Number Values during Solar cycle 24 and Time (In Years) (December 2008 – June 2018)

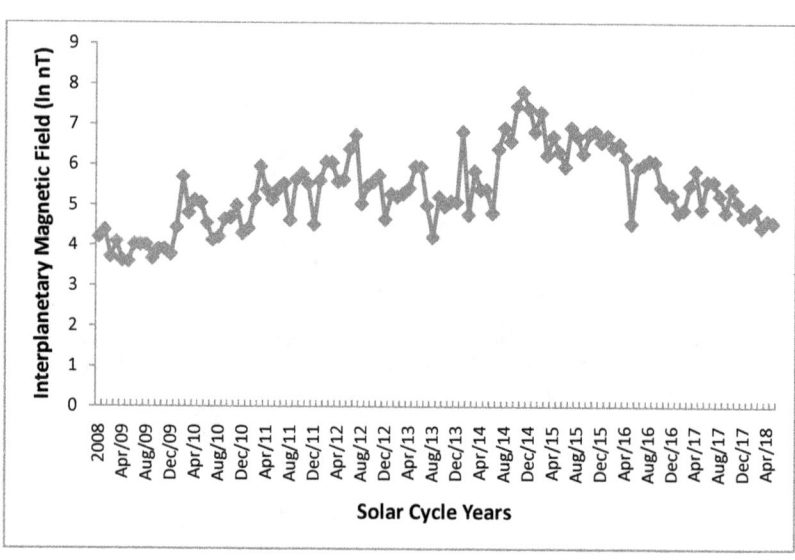

Fig 3.4 Variation between Monthly averaged Interplanetary Magnetic Field Values during Solar cycle 24 and Time (In Years) (December 2008 – June 2018)

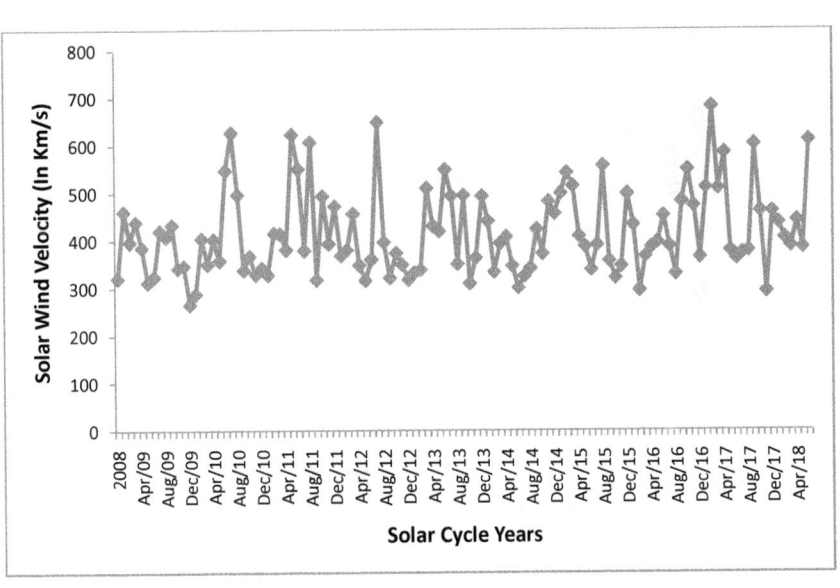

Fig 3.5 Variation between Monthly averaged Solar Wind Velocity Values during Solar cycle 24 and Time (In Years) (December 2008 – June 2018)

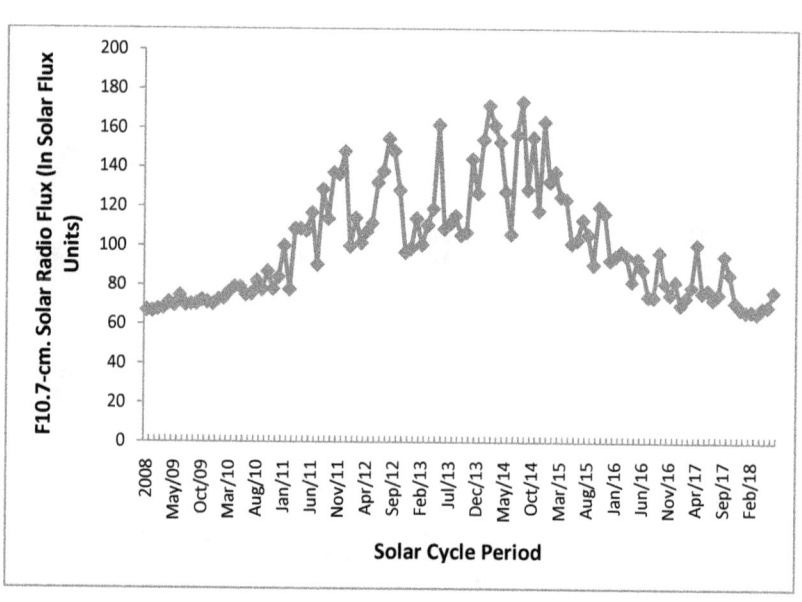

Fig 3.6 Variation between Monthly averaged F10.7 Solar Radio Flux Values during Solar cycle 24 and Time (In Years) (December 2008 –June 2018)

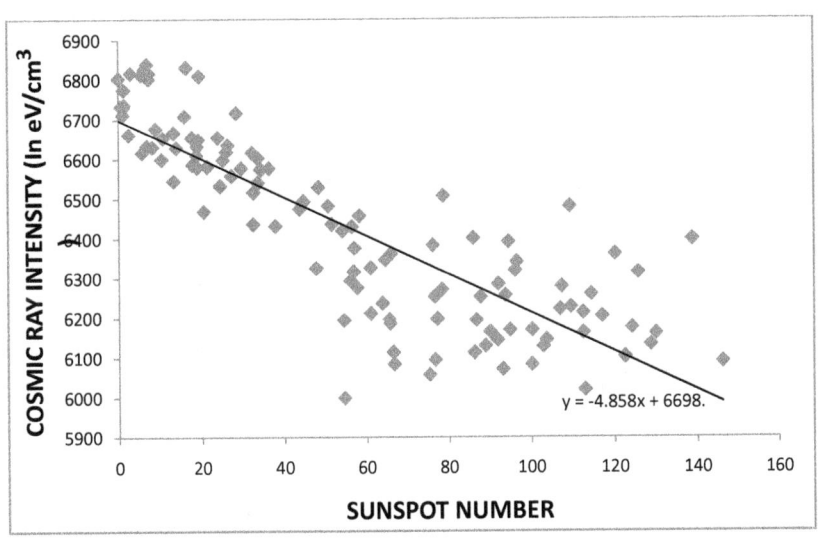

Fig 3.7 Multiple Regression between Cosmic Ray Intensity and SunSpot Number

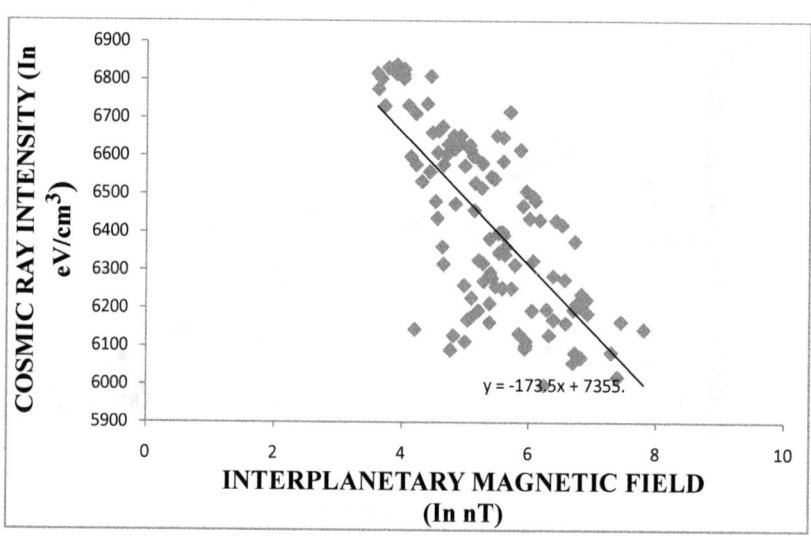

Fig 3.8 Multiple Regression between Cosmic Ray Intensity and Interplanetary Magnetic Field

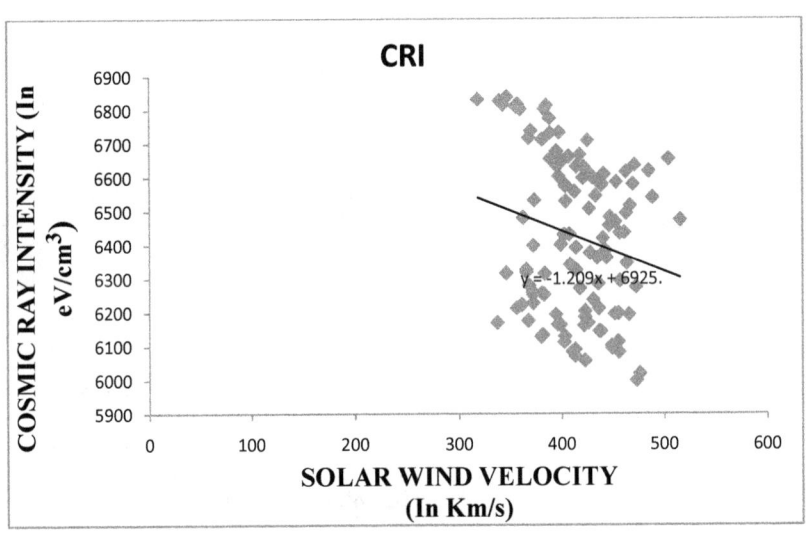

Fig 3.9 Multiple Regression between Cosmic Ray Intensity and Solar Wind Velocity

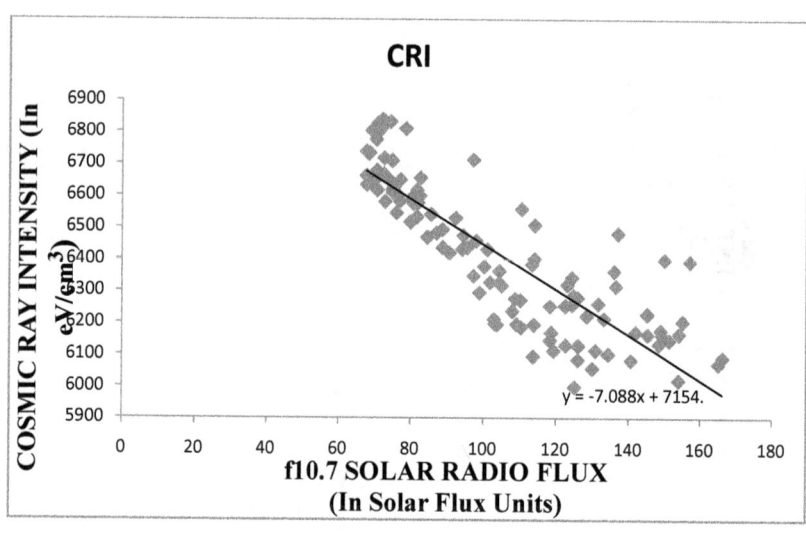

Fig 3.10 Multiple Regression between Cosmic Ray Intensity and F10.7 Sola Radio Flux

3.4 Results and Discussion

The monthly averaged values of CR counts shows steep increase in the first quarter of 2015 and it continues on till 2017. In points where a steep rise in GCR intensity during the declining phase of cycle 24 is observed ,GCR goes 6200 to 6800 counts in the December 2008 to March 2009 epoch. In the fig.2 , SSN values also shows a spotless trend during the descending epoch from December 2008 – April 2009 and October 2009 – December 2009.Observing above results the time variation of Sunspot Number during solar cycle 24 shows a very non-linear trend during the whole cycle.During the ascending epoch, duration January 2009-May 2009 and October 2009-December 2009 were spotless.There has been a steep rise and fall in SSN during the ascending and declining phases with the highest values of SSN achieved during the declining phase.In the duration May 2015-July 2015 and January 2016-March 2016 were also spotless .The time variation of *Interplanetary Magnetic Field (IMF)* during solar cycle 24 shows a steep rise during the initial stage of the ascending epoch and followed regular rise during the whole cycle .Highest value of *interplanetary magnetic field* was obtained in January 2015 ,i.e. in the declining phase .In time variation of *solar wind velocity* during solar cycle 24 shows small fall in the initial stage of ascending epoch accompanied by a regular rise and fall during the ascending and declining phases of cycle. The highest value of *solar wind velocity* has been obtained in the declining epoch in the month of September 2017.The *time* variation of *F10.7 Solar Radio Flux* indicates that the initial phase of ascending epoch were spotless.After that ,it followed a steep rise and fall in the *Solar radio flux*.In contrast to cycle 23 , the cycle 24 also shows an inverse relationship between GCR's and all solar indices for the entire period of investigation .The zenith of the SSN, IMF, F10.7-cm solar radio flux and SWV does not concur precisely with the minimum of the GCR cycle ,which shows that there is a time – lag in the given time series .The solar parameters also shows a secondary apex around the mid 2014 ,compatible with the maxima of cycle 24 which occurred during the declining phase of cycle 24.Only GCR intensity spectra show a minima around the April 2014 and maxima of cycle 24.The regression scatter plots between Cosmic Ray Intensity and Sunspot Number shows negative correlation between them in which Cosmic ray counts is 6850 Counts/min , then Sunspot Number is minimum .The Regression Scatter Plots between

Cosmic Ray Intensity and **Interplanetary Magnetic Field** indicates a negative correlation between them , the scattergram not evenly distributed on two sides of the regression line .These plots between Cosmic Ray Intensity and Solar wind velocity does not show a strict negative correlation between them .It means that Cosmic Ray Intensity and **F10.7 Solar Radio Flux** indicates a negative correlation between them.

3.5 Conclusion

The time variation profile and regression scatter plots between the GCR intensity and various solar parameters advocated the fact that the GCR modulation in the heliosphere occurs mainly due to the three processes of particle diffusion normal to the direction of **heliospheric magnetic field(HMF)**,gradient , curvature ,drift and **heliospheric current sheet(HCS)** effects and convection and **adiabatic energy losses** in the expanding solar wind. The magnetopause which is an uneven line separating the earth's magnetic field and the surrounding plasma is influenced by IMF. The time variation profile of IMF corresponding to cycle 24 disapproved that magnetospheric substorms are initiated by northward turnings of the IMF. A deep insight into the partitioning of solar magnetic field was given which plays a "decisive" physical role in the **acceleration** of the solar wind. High **Solar Wind Velocity** leads to Geomagnetic storms. High **solar wind speed plasma** are deflected eastwards of the Sun and the slow **solar wind speed** plasma are accelerated and deflected westwards of the Sun. This has also been exhibited by the time variation charts of **Solar Wind velocity** . Changes in solar activity also leads to a change in the F10.7 **Solar radio flux**. The F10.7 correlates well with the **sunspot number** as well as a number of Ultra Violet (UV) and visible solar irradiance records. It is measured in units of solar flux units (s.f.u.).The F10.7 can vary from below 50 s.f.u. to above 300 s.f.u. over the course of a solar cycle. The F10.7 Index has proven very valuable in specifying and forecasting space weather.Because it comes from the chromosphere and corona of the sun, it tracks other important emissions that form in the same regions of the solar atmosphere. The Extreme UltraViolet (EUV) emissions that impact the ionosphere and modify the upper atmosphere track well with the F10.7 index. Many Ultra-Violet emissions that affect the stratosphere and ozone also correlate with the F10.7 index. And because this measurement can be made reliably and accurately from the ground in all weather conditions, it is a very robust data set with few gaps or calibration issues.Diffusive shock acceleration is the reason for this uneven scattergram and it has been validated by the power law with a suitable spectral index, and its high efficiency allows SSN's to match the energy requirements for the production of cosmic rays (CR) in our galaxy.CRI is anticorrelated to the **solar wind velocity**. GCRs respond more strongly to the A < 0 magnetic polarity state. The arrival of GCR is delayed by a few solar rotations during solar minima (A < 0) against the outward convection of solar wind carrying the IMF lines .The recovery for the odd cycles replenish to a higher level than the even cycles.The charged particle drifts in inhomogeneous **interplanetary magnetic field** was studied and it was specified that they are the key factors for the aforesaid effects. Highest CRI does not have any correlation with R_z and f10.7 Solar Radio Flux, but Lowest CRI has good correlation.This could be explained by the fact that the rise in CRI across the solar activity minima is due to the weak detraction in the CRI,while the fall in LCRI across the solar activity maxima is due to the strong detraction in CRI .The outstanding question is how much merging occurs and what happens with these diffusion barriers in the heliosheath. These aspects are currently being investigated.F10.7 radio emissions originates high in the chromospheres and low in the corona of the solar atmosphere.

CHAPTER 4
VARIATIONS IN COSMIC RAYS WITH THE 11-YEARS SOLAR ACTIVITY

4.1 Introduction

Solar Cycle can also be defined as the duration in which the solar magnetic flux rises up to the Sun's surface. Durations when all the indicators of solar activity are maximum are called Solar Maximum or Solar Maxima and durations when all the indicators of solar activity are minimum are called Solar Minimum or Solar Minima. Cosmic ray flux is tremendously affected by the 11-years solar activity. The high-energy cosmic rays are dispersed by the external evolution of solar eject into interplanetary space. Durham (2006) explained that solar activity has short –term recurrences like 5-year, quasi-biennial(2-3 year) ,quasi-triennial(3-4 year) and a 27-day recurrence in addition to the 11-year sunspot number cycle and 22-year magnetic cycle and probably an 80-year cycle as the most prominent recurrence, 1.7, 1.3, 1.0 year, a few (5–7) months, have also been reported. The outward emanating cosmic ray flux in the interplanetary space is anticorrelated with the level of solar cycle. The galactic cosmic rays make a journey towards the Earth and their solar transition of the intensity of, as a consequence of the changes of topology of the interplanetary magnetic field in the course of the solar cycle, is related to the solar corona structure and activity. An anti-correlation exists between cosmic ray intensity and solar activity .The 11-years solar activity is predominantly indicated by the sunspot activity .Contradictory to the times of high solar activity, the sunspot cycle during MM was, contrary, influenced by a 22-year cycle. Since the measurement of cosmic rays began in the early 20th century, the CR flux has been the most uncommon during solar cycle 20.

On observing both high and low solar activity periods, the cosmic ray flux has been found to be different. The epochs of increased cosmic ray flux have been found during the interval of the maximum of the 11-year solar cycle, when the flare and eruptive activity of the Sun increases suddenly. During these periods, two or three powerful and fast evolving active regions (AR) produce many flares and coronal mass ejections

(CMEs) and provide the most powerful energy release in the cycle. These epochs occurred during the first years of the descending phase of the solar cycle ,during November 1960 (cycle 19), August 1972 (cycle 20), June–July 1982 (cycle 21), August 1991 (cycle 22), and March–April 2001 (cycle 23) and during December 2008-April 2014 (cycle 24). During descending phase of the solar cycle, the solar wind velocity also increases .This happens due to a complex solar magnetic structure whose existence was recognized by Gnevyshev , who suggested that the 11-year cycle does not contain one but two waves of activity with different physical properties. The periods of reduced solar activity between the two peak waves during the solar maximum phase were named Gnevyshev Gaps (GG) and even now the nature of the complicated double-peak structure of the solar cycle maximum is still obscure.

4.2 Main Factors Affecting Variations in Cosmic Rays

There are three main factors affecting cosmic ray variations

1. Sunspot Number

Sunspots are dark spots on the sun where intense magnetic field loops up through the surface from the interior with flux densities of 0.1 to 0.4T, while the average solar magnetic field flux density is 0.001T. **Sunspots** are regions where the **solar** magnetic field is very strong. The sun has been repeatedly recorded for observing sunspots when the first telescope emerged in 1609. The notion of compiling the information about the sunspot number from various bystanders was initiated by a Swiss scientist Rudolf Wolf in 1848 . Sunpot Number is also commonly known as Wolf Number or International Sunspot Number or Relative Sunspot Number or Zurich Number.Sunpot Number is still being produced today at the observatory of Brussels. As compared to the adjoining regions having temperatures of about 5778K ,a typical sunspot have temperatures of about 3700K and so they seem to be dark on the photosphere. Their diameters are typically less than or equal to 50000 km. Their lifetimes are on the order of a few days to weeks. Birkeland (1916) pointed out that the origination points of sunspots are the currents of plasma deep in the convective zone that flow to the photosphere due to the differential rotation of sun. They appear in pairs with opposite magnetic polarity.

Sunspots appear in pairs generally within 5 deg of the solar equator.Sunspot number's first evidences was available from China over 2000 years ago . Sammuel Heinrich Schwabe, a German astronomer discovered the sunspot number cycles in 1843 through solar observation.

a. Relation of Sunspot Number with Solar Cycle

Solar maximum and solar minimum are the periods of maximum and minimum sunspot counts hence,sunspot numbers are also called sunspot cycles. Cycles span from one minimum to the next. Scrutinization of sunspots started from the early 17th century and the sunspot time series is the longest, continuously observed (recorded) time series of any natural phenomena.Sun has a dipolar magnetic field revolving every 11 years ,accompanying the 11 year quasi-periodicity in sunspots however, the peak in the dipolar field lags the peak in the sunspot number, with the former occurring at the minimum between two cycles.Solar activity is managed both by the sunspot cycle and transient aperiodic processes by governing the environment of the Solar System planets by creating space weather and impact space- and ground-based technologies as well as the Earth's atmosphere and also possibly climatic fluctuations .The sunspot cycle is rather irregular. At the advent of odd cycles than for even cycles ,sunspot number corresponds to a lower (inverted) cosmic-ray count rate; conversely, on the decay of odd-numbered cycles, sunspot number correlates to a higher (inverted) count rate than for even cycles. Sunspot cycle is related to solar cycle in a way that higher solar activity means higher sunspot cycle.During maximum phase of a solar cycle some crucial terrestrial consequences occurs such as the higher solar emission of extreme-ultraviolet and ultraviolet flux resulting in a transition of middle and upper terrestrial atmosphere, and total solar irradiance, affecting the terrestrial climate as well as the coronal mass ejection and interplanetary shock rates. There is also a high probability for the occurrence of large solar flares, with associated energetic solar particles causing phenomena such as communications disturbances, failures in electronic solid state components, etc.Solar cycles 17 – 23 have encompassed a distinctively strong activity called the "Modern Maximum" .Cycles 5, 6, and 7 have been abruptly weak, forming the so-called "Dalton Minimum".Cycles 12 – 16 have been moderately weak called as the "Gleissberg

Minimum". In addition, the 11-year solar cycle has an asymmetric shape with a shorter ascending (≈ 4-year on an average) and a longer (≈ 7-year) descending phase. The asymmetry is typically larger for shorter cycles. Expulsion of solar material, extent of solar radiation, sunspot's number and size, solar flares, and coronal loops all reveal a synchronized fluctuation, from active to quiet to active again, with a period of 11 years. This cycle has been observed for centuries by changes in the Sun's appearance and by terrestrial phenomena such as auroras.

b. **Relative Sunspot Number**

Rudolf Wolf who was a scientist in Zurich Observatory invented a method to compute the daily sunspot number from the observed number of sunspot groups. Observations of Wolf were primarily based on the observations carried out by Heinrich Schwabe on sunspot number cycles. Wolf retrieved sunspot number data from 1610 onwards by taking the duration of the sunspot number cycle as 11.1 years.

Wolf used the following Equation to compute the Sunspot Number :

$$R = k(10g + s) \qquad (1)$$

where, R is the sunspot number,

k is a scaling factor depending on the specific observatory, generally ≤ 1,

g is the number of sunspot groups, and

s is the number of sunspots.

The scaling factor k describes the observing characteristics, such as types of telescope and observing conditions. Sunspot numbers from a number of international observatories are then combined to arrive at a daily value from which weekly, monthly, and yearly averages are computed. Archived direct observations of the sunspot numbers are publically available from the year 1749 onwards and their authenticity have also been established. International sunspot numbers have been made available on a daily basis from 1849 onwards. Rudolf Wolf was the primary observatory from 1848 to 1893 and used correction factor k = 1.0. The Swiss Federal Observatory was the nodal agency to provide sunspot numbers upto 1980. From 1981 onwards, the Royal Observatory of Belgium with Koeckelenbergh as the primary observer has been providing the international sunspot number. Today, this is available through Solar Influences Data Analysis Center(SIDC), of Royal Observatory, Belgium.There are many other observers across the globe and whose observations are used when the primary observers data are not available. International sunspot number is considered as the key indicator of solar activity.

Generally, sunspot numbers are available in four different formats or categories viz.

1. Daily number,

2. Monthly average,

3. Yearly average, and

4. Smoothed sunspot number formats.

Today various additional form of sunspot numbers exist .

They are:

1. The Boulder sunspot number,

2. American sunspot number,

3. Group sunspot number,

c. Boulder Sunspot Number:

The Boulder sunspot number are acquired from the daily solar region summary published by the government agency, US Air Force and National Oceanic and Atmospheric Administration (USAF/NOAA). This encapsulation is being attained from the sunspot drawings captured from the Solar Optical Observing Network (SOON) sites from 1977 onwards. Then, the Boulder sunspot number is obtained using Equation (1.2) with a correction factor of k = 1.0. The magnitude of the Boulder sunspot number is generally

55% larger than the international sunspot number (correction factor k = 0.65) . This data is available on a daily basis.

d. American Sunspot Number

American sunspot number has been produced by the American Association of Variable Star Observers (AAVSO). These sunspot numbers are available from the year 1944 onwards to the present day. The American number often deviates from the international sunspot number. The American sunspot number is generally 3% lower than the international sunspot number.

e. Group Sunspot Number, R_G:

There was a duration before the year 1850 when Wolf sunspot number time series was of poor quality and a period before before 1750 when Wolf sunspot number time series was less reliable. Due to this, there was a need to re-evaluate early sunspot data. Hoyt et.al performed complete ancient search on the Wolf sunspot numbers and produced a new time series of sunspot activity, called as the group sunspot numbers. This index indicates the number of sunspot groups. Generally this data is being computed as the average of the observations collected from multiple observers (rather than using the primary/secondary/tertiary observer system). Then these averaged values are normalized to the international sunspot numbers using Equation (2),

$$R_G = \frac{12.08}{N} \sum_{i=1}^{N} k_i G_i \qquad (2)$$

where N is the number of observers, k_i is the i th observer's correction factor, G_i is the number of sunspot groups observed by ith observer, and 12.08 is the normalizing factor to the international sunspot number.Hathaway et.al, had shown that the group sunspot numbers follows the international sunspot number very closely. Group sunspot numbers are not directly available after 1995. The major application of group sunspot number is in reconstructing the sunspot number observations back to the year 1610. These sunspot numbers are available from the website of National Oceanic and Atmospheric Administration (NOAA). The international sunspot numbers are available from the website of SIDC.

2. Sunspot Areas

Direct evaluators of solar activity are the Sunspot areas. Sunspot areas and positions were recorded by the Royal Observatory, Greenwich (RGO) from May of 1874 to the end of 1976. They used the measurements of photographic plates obtained from various observatories in Cape Town, South Africa; Kodaikanal, India; and Mauritius.Along with sunspot numbers, another measurement, which depicts the solar activity in a more quantified way is the F10.7cm solar index. It has direct application in the satellite orbit prediction studies.This index has a very high correlation with the sunspot numbers.

3. Solar Wind Velocity

A jet of charged particles released from the upper ambience, called the corona is called the Solar Wind. This plasma mostly consists of electrons, protons and alpha particles with kinetic energy between 0.5 and 10 keV. The constitution of the solar wind plasma also includes a mixture of materials found in the solar plasma: sign amounts of heavy ions and atomic nuclei C, N, O, Ne, Mg, Si, S, and Fe. There are also small traces of some other nuclei and isotopes such as P, Ti, Cr, Ni, Fe 54 and 56, and Ni 58,60,62. interplanetary magnetic field is implanted within the solar-wind plasma. The density, temperature and speed of solar wind was studied and found that it undergoes a change with time and with different solar latitude and longitude. The velocity of a solar wind is called Solar Wind Velocity or Flow Speed.At a distance of more than a few solar radii from the Sun, the solar wind reaches speeds of 250 to 750 kilometers per second and is supersonic,i.e. it moves faster than the speed of the fast magnetosonic wave, which is a wave driven by both pressure (thermal and magnetic) and magnetic tension. There are two types of magnetosonic waves, the *fast* magnetosonic wave and the *slow* magnetosonic wave. Both fast and slow magnetosonic waves have been recently discovered in the solar corona.

4.3 Data Analysis

The variation of Sunspot number and Solar wind velocity with cosmic ray intensity on 13 high amplitude events and 7 low amplitude events during the low solar activity epoch ,i.e. from December 2008 to April 2014 have been discussed in this chapter. Solar cycle 24 began in 2008 and ended in the middle of 2019. The daily averaged values of the Solar Wind Velocity and Sunspot Number have been taken from the Omniweb Website of the Goddard Space Flight Centre ,NASA .This website uses data obtained from the IMP 8, Geotail, Wind and ACE spacecrafts.

4.4 Results

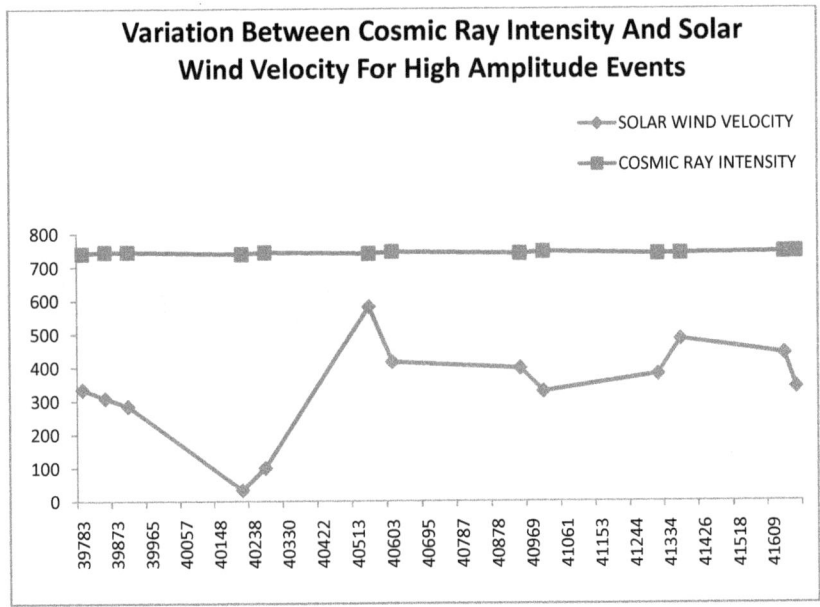

Fig 4.1 Variation Between Cosmic Ray Intensity and Solar Wind Velocity for High Amplitude Events

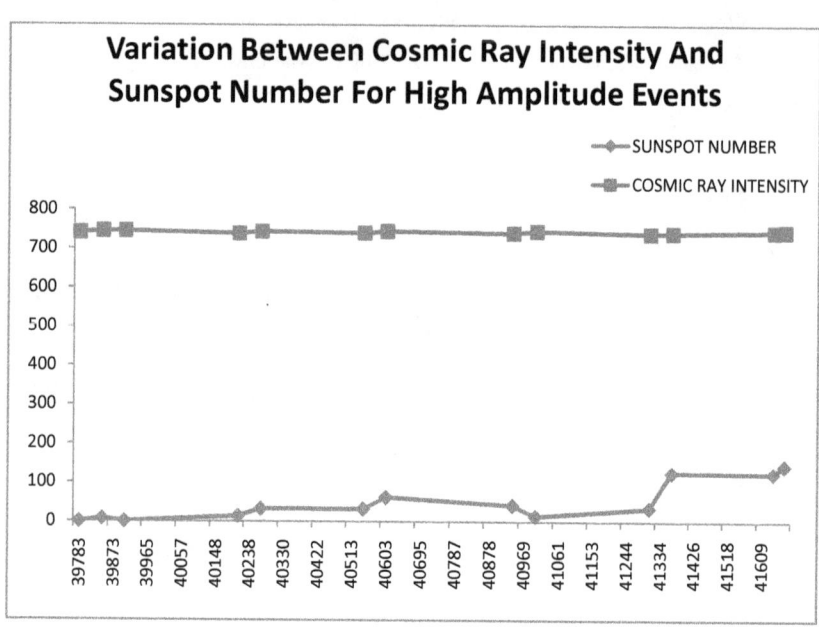

Fig 4.2 Variation Between Cosmic Ray Intensity And Sunspot Number For High Amplitude Events

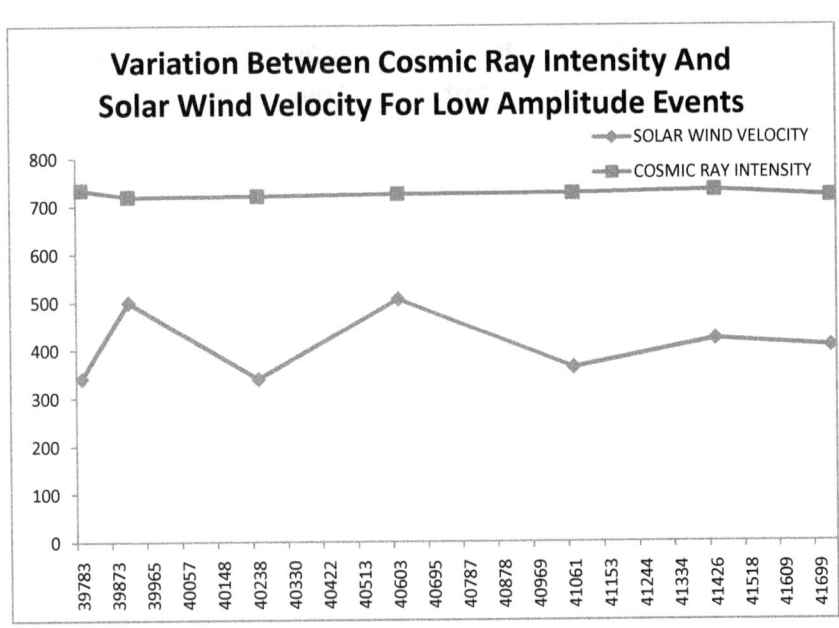

Fig 4.3 Variation Between Cosmic Ray Intensity and Solar Wind Velocity for Low Amplitude Events

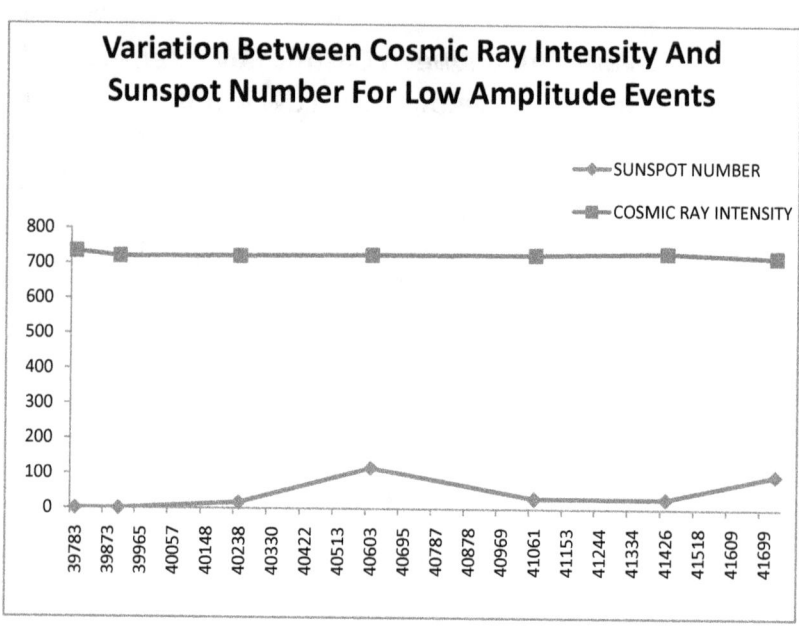

Fig 4.4 Variation Between Cosmic Ray Intensity and Sunspot Number for Low Amplitude Events

4.5 Discussion

Below plot shows the time series of Solar Wind Velocity and Cosmic ray intensity with different high amplitude events during the descending phase of solar Cycle 24 ,i.e. from December 2008-April 2014.The values of solar wind velocity shows a linear decrease during the first three values, whereas the values of cosmic ray intensity shows a simultaneous linear increase during the first three values. During the next two values, the solar wind velocity values shows a steep increase, whereas the cosmic ray intensity values shows a small linear increase.. During the next two values ,t he solar wind velocity values shows a small steep decrease while the cosmic ray intensity values shows a corresponding steep increase. During the next two values, the solar wind velocity value shows a steep decrease while the cosmic ray intensity values show a small steep increase. The next two values of solar wind velocity witness a steep increase while the cosmic ray intensity value shows a small increase. .Last two values of solar wind velocity witness a steep decrease while the cosmic ray intensity values show a small increase. Above plot shows the time series of Sunspot Number and Cosmic ray intensity with different high amplitude events during the descending phase of solar Cycle 24, i.e. from December 2008-April 2014.The values of Sunspot Number shows a simultaneous linear increase and decrease during the first three values, whereas the values of cosmic ray intensity shows a simultaneous small linear increase and decrease during the first three values. During the next two values, the sunspot number values show a minute steep increase, whereas the cosmic ray intensity values show a small linear increase. During the next two values, the sunspot number values show a small steep decrease while the cosmic ray intensity values shows a corresponding minute steep increase. During the next two values, the sunspot number values show a steep decrease while the cosmic ray intensity values shows a small steep increase. The next two values of sunspot number witnesses a steep increase while the cosmic ray intensity values show a small increase. Last two values of sunspot number witnesses a small increase while the cosmic ray intensity values shows constancy.

Below plot shows the time series of Solar Wind Velocity and Cosmic ray intensity with different low amplitude events during the descending phase of solar Cycle 24 ,i.e. from December 2008-April 2014.The values of Solar Wind Velocity shows a steep increase during the first two values, whereas the values of cosmic ray intensity shows a small linear increase during the first two values.During the next two values ,the solar wind velocity values shows a steep decrease,whereas the cosmic ray intensity values shows a small linear increase.During the next two values ,the solar wind velocity values shows a small steep increase while the cosmic ray intensity values shows a corresponding small linear increase.During the next two values ,the solar wind velocity values shows a steep decrease while the cosmic ray intensity values shows a small linear increase. The next two values of solar wind velocity witnesses a steep increase while the cosmic ray intensity values shows a small linear increase.Last two values of solar wind velocity witnesses a small linear decrease while the cosmic ray intensity values also shows a corresponding linear decrease.

Below plot shows the time series of Sunspot Number and Cosmic ray intensity with different low amplitude events during the descending phase of solar Cycle 24 ,i.e. from December 2008-April 2014.The values of Sunspot Number shows constancy during the first two values, whereas the values of cosmic ray intensity shows a small linear increase during the first two values.During the next two values ,the sunspot number values shows a small linear increase,whereas the cosmic ray intensity values shows corresponding constancy.During the next two values ,the sunspot number values shows a steep increase while the cosmic ray intensity values shows a corresponding small linear increase. During the next two values ,the sunspot number values shows a steep decrease while the cosmic ray intensity values shows constancy. The next two values of sunspot number witnesses constancy while the cosmic ray intensity values shows a corresponding constancy.Last two values of sunspot number witnesses a steep increase while the cosmic ray intensity values also shows a corresponding small linear decrease.

4.6 Conclusion

1. Among various indicators of solar activity , Sunspot number is the main parameter whose variation leads to changes in cosmic ray intensity.

2. *Sunspot numbers* leads the *cosmic ray intensity* and there is a slight phase coherence *at* periodicity around 8 days *at* the descending phase of *cycle 23* from 2004 to 2006.

CHAPTER 5
STUDY OF EFFECT OF SOLAR ACTIVITY ON EARTH'S ATMOSPHERIC WEATHER

5.1 Introduction

Sun's interior is a source of many complex dynamical phenomena, which pave a way for new discovery and knowledge. Surface temperature of sun ranges from 5000-6000K. Sun's mass comprises of hydrogen about 74%, helium about 24%, and all other elements together about 2%. The age of the Sun is about 4.5 billion years old. Currently it has completed halfway its development. Many observers noticed intermediate changes happening in the sun in due course. Based on the physical characteristics, the Sun is divided into six regions,

Core

Sun's innermost region having a radius of about 0.25 solar radius(174000 km) is termed as Core. Its temperature is about 16,000,000 K. Core density is 150 times greater than that of water (150,000 kg/m^3). The high temperature and density of the core ionizes all constituent elements and sustain thermonuclear fusion which produces excess energy. Helium nuclei are produced from hydrogen nuclei by the nuclear fusion process. This process converts four hydrogen nuclei of mass 4 × (1.67262171 × 10^{-27}kg) = 6.69048684 × 10^{-27}kg into one helium nucleus with a mass of 6.64465598 × 10^{-27}kg [56]. This loss of 0.66% of the initial mass m is being converted into energy E. This is proved by Einstein's formula,

$$E = mc^2 \quad (1)$$

where, c is the velocity of light. Enormous amount of energy is being produced in the core of the Sun through this way.

Radiative zone

The radiative zone consists of highly ionized gas and ranges from about 0.25 solar radius(174,000 km) to 0.75 solar radius(522,000 km). Energy delivery takes place from the core to the outer surface of the irradiative zone mainly by gamma ray photon diffusion process.

Convection zone

Sun's exterior region is the convection zone. It ranges from about 0.75 solar radius to the visible surface of the Sun. The temperature in this zone varies from about 2, 000, 000K at the bottom to about 5778K at the surface of the convection zone. At the Surface density is about $2\times10^{-4} kgm^{-3}$. Convection zone's surface acts like a radiation emitter for the sun. The convection zone comprises of about 70% hydrogen and 28% helium and 2% of carbon, nitrogen, and oxygen. The convection zone consists of plasma and hence, unlike a solid body, Sun's surface rotates differentially with its equatorial region rotating with a duration of 26 days and the polar region rotating with a duration of 32 days.

Photosphere

Straight visible surface of the sun having a thickness of about 100 to 500 km is termed as Photosphere. Its temperature is about 5778K. The particle density of the photosphere is about 1% of the density of Earth atmosphere at sea level. Photosphere emanates almost all of the radiation emitted by the sun.

Chromosphere

The Chromosphere is the region extending from 2000 to 5000 km above the photosphere. The density of the chromosphere is very low, it being only 104 times that of the photosphere. The density of the chromosphere decreases with distance from the center of the sun.

Corona

Sun's exterior plasma atmosphere is termed as Corona. This can be seen during a total solar eclipse and has no well defined outer surface. Its temperature ranges from about 0.5×10^6 to 2×10^6K. The plasma particle density is about 10^{11} particles per unit area.

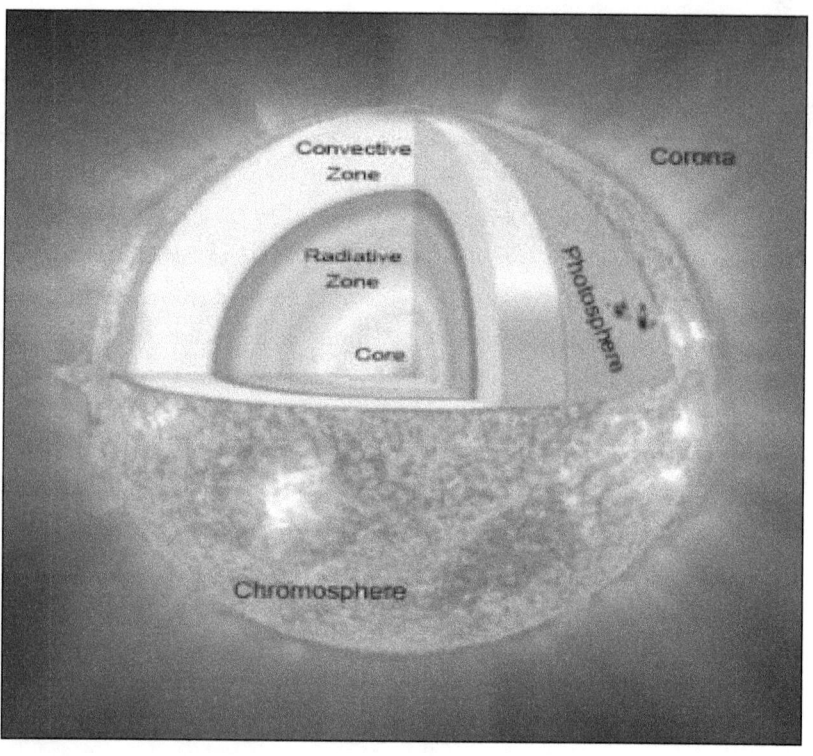

Fig 5.1 Schematic of the six divisions of Sun {Website: https://scied.ucar.edu/sun-regions}

5.2 Sunspots

Sunspots are dark spots on the sun where intense magnetic field loops up through the surface from the interior with flux densities of 0.1 to 0.4T, while the average solar magnetic field flux density is 0.001T. **Sunspots** are regions where the **solar** magnetic field is very strong. The sun has been repeatedly recorded for observing sunspots when the first telescope emerged yuyuin 1609. Carroll & Ostlie (1995) studied that the notion of compiling the information about the sunspot number from various bystanders was initiated by a Swiss scientist **Rudolf Wolf** in 1848. Schrijver & Zwaan (2000) pointed out that Sunpot Number is also commonly known as Wolf Number or International Sunspot Number or Relative Sunspot Number or Zurich Number.Sunpot Number is still being produced today at the observatory of Brussels. As compared to the adjoining regions having temperatures of about 5778K ,a typical sunspot have temperatures of about 3700K and so they seem to be dark on the photosphere. Their diameters are typically less than or equal to 50000 km. Their lifetimes are on the order of a few days to weeks. The origination points of sunspots are the currents of plasma deep in the convective zone that flow to the photosphere due to the differential rotation of sun. They appear in pairs with opposite magnetic polarity. Sunspots appear in pairs generally within 5 deg of the solar equator.Sunspot number's first evidences was available from China over 2000 years ago . Sammuel Heinrich Schwabe, a German astronomer discovered the sunspot number cycles in 1843 through solar observation.

5.3 SOLAR ACTIVITY

Solar activity is a collection of different momentary phenomena taking place in the solar plasma . From the initial phase of the early 17th century, scientists knew about the emergence and fading away of black spots on the visible surface of the Sun.Scientific fraternity instituted in the 19th century that, solar activity varies in an 11-year time period. Significant observation of solar activity variation shows that the solar activity progresses from a minimum value and reaches to a maximum value, and then, again comes back to a minimum value.Two successive solar activity minima are separated by about 11 year. Prominent solar activity indicators are sunspots, radio flux expulsions, granules, super granules, spicules, faculae in the photosphere, flares and plages in the chromospheres, prominence, coronal structure, and mass ejections in the corona ,predominant among them are sunspots and F10.7 Solar radio flux.Solar activity exhibits a high correlation with sunspot number and F10.7cm flux density. These parameters are low during minimum solar activity and high during maximum solar activity.

5.3 Solar Cycle

Different momentary phenomena taking place in the solar plasma which varies in a duration of 11 years is called as Solar Cycle. Durations when all the indicators of solar activity are maximum are called Solar Maximum or Solar Maxima and durations when all the indicators of Solar activity are minimum are called Solar Minimum or Solar Minima. Solar Cycle can also be defined as the duration in which the solar magnetic flux rises up to the Sun's surface.

5.4 Relation of Sunspot Number with Solar Cycle

Solar maximum and solar minimum are the periods of maximum and minimum sunspot counts hence, sunspot numbers are also called sunspot cycles. Cycles span from one minimum to the next. Scrutinization of sunspots started from the early 17th century and the sunspot time series is the longest, continuously observed (recorded) time series of any natural phenomena. Sun has a dipolar magnetic field revolving every 11 years, accompanying the 11 year quasi-periodicity in sunspots however, the peak in the dipolar field lags the peak in the sunspot number, with the former occurring at the minimum between two cycles. Geiss et al. (1995) enunciated that Solar activity is managed both by the sunspot cycle and transient a periodic processes by governing the environment of the Solar System planets by creating space weather and impact space- and ground-based technologies as well as the Earth's atmosphere and also possibly climatic fluctuations .The sunspot cycle is rather irregular. Solar cycles 17 – 23 have encompassed a distinctively strong activity called the "Modern Maximum" .Cycles 5, 6, and 7 have been abruptly weak, forming the so-called "Dalton Minimum".Cycles 12 – 16 have been moderately weak called as the "Gleissberg Minimum". In addition, the 11-year solar cycle has an asymmetric shape with a shorter ascending (≈ 4-year on an average) and a longer (≈ 7-year) descending phase. The asymmetry is typically larger for shorter cycles. Expulsion of solar material, extent of solar radiation ,sunspot's number and size , solar flares, and coronal loops all reveal a synchronized fluctuation, from active to quiet to active again, with a period of 11 years. This cycle has been observed for centuries by changes in the Sun's appearance and by terrestrial phenomena such as auroras.

5.5 Relative Sunspot Number

Rudolf Wolf who was a scientist in Zurich Observatory invented a method to compute the daily sunspot number from the observed number of sunspot groups.Observations of Wolf were primarily based on the observations carried out by Heinrich Schwabe on sunspot number cycles. Wolf retrieved sunspot number data from 1610 onwards by taking the duration of the sunspot number cycle as 11.1 years.

Wolf used the following Equation to compute the Sunspot Number :

$$R = k(10g + s) \qquad (2)$$

where, R is the sunspot number,

k is a scaling factor depending on the specific observatory, generally ≤ 1, g is the number of sunspot groups, ands is the number of sunspots. The scaling factor k describes the observing characteristics, such as types of telescope and observing conditions. Sunspot numbers from a number of international observatories are then combined to arrive at a daily value from which weekly, monthly, and yearly averages are computed. Archived direct observations of the sunspot numbers are publically available from the year 1749 onwards and their authenticity have also been established. International sunspot numbers have been made available on a daily basis from 1849 onwards. Rudolf Wolf was the primary observatory from 1848 to 1893 and used correction factor k = 1.0.

The primary observatories from 1749 are listed here:

1. Staudacher (from 1749 to 1787),

2. Flaugergues (from 1788 to 1825)

3. Schwabe (from 1826 to 1847)

4. Wolf (from 1848 to 1893)

5. Wolfer (from 1893 to 1928)

6. Brunner (from 1929 to 1944)

7. Waldmeier (from 1945 to 1980)

The Swiss Federal Observatory was the nodal agency to provide sunspot numbers upto 1980. From 1981 onwards, the Royal Observatory of Belgium with Koeckelenbergh as the primary observer has been providing the international sunspot number. Today, this is available through Solar Influences Data Analysis Center(SIDC), of Royal Observatory, Belgium.There are many other observers across the globe and whose observations are used when the primary observers data are not available. International sunspot number is considered as the key indicator of solar activity.Generally, sunspot numbers are available in four different formats or categories viz.

1. Daily number,

2. Monthly average,

3. Yearly average, and

4. Smoothed sunspot number formats.

Today various additional form of sunspot numbers exist .

They are:

1. The Boulder sunspot number,

2. American sunspot number,

3. Group sunspot number,

The Boulder sunspot number:

The Boulder sunspot number are acquired from the daily solar region summary published by the government agency, US Air Force and National Oceanic and Atmospheric Administration (USAF/NOAA). This encapsulation is being attained from the sunspot drawings captured from the Solar Optical Observing Network (SOON) sites from 1977 onwards. Then, the Boulder sunspot number is obtained using Equation (5.2) with a correction factor of k = 1.0. The magnitude of the Boulder sunspot number is generally 55% larger than the international sunspot number (correction factor k = 0.65) . This data is available on a daily basis.

American sunspot number:

American sunspot number has been produced by the American Association of Variable Star Observers (AAVSO). These sunspot numbers are available from the year 1944 onwards to the present day. The American number often deviates from the international sunspot number. The American sunspot number is generally 3% lower than the international sunspot number.

Group Sunspot Number, R_G:

There was a duration before the year 1850 when Wolf sunspot number time series was of poor quality and a period before before 1750 when Wolf sunspot number time series was less reliable. Due to this, there was a need to re-evaluate early sunspot data. Hoyt et.al performed complete ancient search on the Wolf sunspot numbers and produced a new time series of sunspot activity, called as the group sunspot numbers. This index indicates the number of sunspot groups. Generally this data is being computed as the average of the observations collected from multiple observers (rather than using the primary/secondary/tertiary observer system). Then these averaged values are normalized to the international sunspot numbers using Equation (3),

$$R_G = \frac{12.08}{N} \sum_{i=1}^{N} k_i G_i \quad (3)$$

,where N is the number of observers, k_i is the i th observer's correction factor, G_i is the number of sunspot groups observed by ith observer, and 12.08 is the normalizing factor to the international sunspot number.Hathaway et.al, had shown that the group sunspot numbers follows the international sunspot number very closely. Group sunspot numbers are not directly available after 1995. The major application of group sunspot number is in reconstructing the sunspot number observations back to the year 1610. These sunspot numbers are available from the website of National Oceanic and Atmospheric Administration (NOAA). The international sunspot numbers are available from the website of SIDC.

5.6 Sunspot Areas

Direct evaluators of solar activity are the Sunspot areas. Sunspot areas and positions were recorded by the Royal Observatory, Greenwich (RGO) from May of 1874 to the end of 1976. They used the measurements of photographic plates obtained from various observatories in Cape Town, South Africa; Kodaikanal, India; and Mauritius. Along with sunspot numbers, another measurement, which depicts the solar activity in a more quantified way is the F10.7cm solar index. It has direct application in the satellite orbit prediction studies. This index has a very high correlation with the sunspot numbers.

5.7 Granules and super granules

Small lamellar biological regions of 400 to 1000 km encompassing the photosphere are termed as granules. Granules are generally convection regions of hot plasma arising from the bottom of the convection zone and generally remains for 5 to 20 minutes on the photosphere before cooling. Super granules are larger renditions of granules with sizes of 35,000 km.

Spicules

Cascades of hot plasma having 500 km diameter and extending from 1000 to 10000 km above the photosphere into the chromospheres are termed as Spicules. They travel at a speed up to 22 km/s.

Faculae

Glittering powdery structure on the sun's photosphere that are hotter than the encircling plasma are termed as faculae. Faculae make their appearance some hours before sunspots appear in the same location and can remain there for few months after the sunspots disappear. They are smaller than sunspots.

Plages

Plages are Shining regions that encircle sunspots representing regions of higher temperature and density within the chromospheres are termed as plages. They appear with sunspots but last longer.

Prominence

Curves of Plasma extending from the photosphere into the corona and back are termed as Prominence. They are cooler and denser than the adjoining corona and stay for weeks and sometimes months suspended by constrained magnetic fields.

Flares

Sunspots are sometimes followed by sudden and intense eruptions in the corona, termed solar flares, spiraling outward and emanating high energy particles and radiation in a broad spectrum from radio waves to gamma rays. Special lifetimes of solar flares are 1 hours to 2 hours.

5.8 Coronal Mass Ejection (CME)

Bigger amounts of plasma with an organized magnetic field that emanate from the sun in several hours are called Coronal mass ejections(CME). Suess (1999) explained that ejections of large quantities of matter and electromagnetic radiation takes place into space above the Sun's surface in coronal mass ejections, either near the corona, or farther into the planetary system, or beyond. When it occurs near the corona ,it is sometimes termed as Solar Prominence ,and when it occurs beyond the planetary system, it is called interplanetary CME .The ejected material is a magnetized plasma consisting primarily of electrons and protons. Coronal mass ejections are associated with enormous changes and disturbances in the coronal magnetic field. They are usually observed with a white-light coronagraph.

Physical Properties of Coronal Mass Ejections

CME's are connected with the solar flares and prominences. They move with a velocity of 200 km/s to as fast as near 3000 km/s based on SOHO/LASCO measurements between 1996 and 2003. These speeds correspond to transit times from the Sun out to the mean radius of Earth's orbit of about 13 hours to 86 days (extremes), with about 3.5 days as the average. Those CME's having speeds in the range of 3000 km/s are termed as fast CME's and **can** reach our planet in as little as 15-18 hours. Slower **CMEs can** take many days to arrive. Fast CMEs are very often associated with strong geomagnetic storms (Srivastava & Venkatakrishnan 2004, Yurchyshyn et al. 2004, 2005) and the correlation is the best when an earthward CME is associated with a magnetic cloud (Gopalswamy 2010). They have plasma masses of electrons and protons up to 2×10^{13} kg. On ejecting from the Sun, high speed particles and strong magnetic fields can hurl earthward thus causing a significant impact on the near Earth space environment such as adverse effects on satellites and communications, electric power, pipelines, etc. ,and on reaching earth, their impact is tremendous causing geomagnetic storms, damaging power grids and spacecraft . The increased radiation environment can damage electronic equipments and reveal humans to excessive radiation.Coronal mass ejections has a vast impact on solar activity. Earliest speculation and critical study of coronal mass ejections are very much needed in the present day, since human beings are highly dependent on space related technologies for day to day life activities.Zirker (1977) explained that a typical coronal mass ejection may cause three distinguishing characteristics: a cavity of low electron density, a dense core (the prominence, which is a shining area embedded in this cavity), and a bright leading edge.Most CME's originate from active regions on the Sun's surface, such as groupings of sunspots associated with frequent flares. These regions have closed magnetic field lines, in which the magnetic field strength is large enough to contain the plasma. These field lines must be broken or weakened for the ejection to escape from the Sun.CMEs may also be launched in quiet surface regions. During solar minimum, CMEs form primarily in the coronal streamer belt near the solar magnetic equator. During solar maximum, they originate from active regions whose latitudinal distribution is more homogeneous.The frequency of ejections depends on the phase of the solar cycle: from about 0.2 per day near the solar minimum to 3.5 per day near the solar maximum. The phenomenon of magnetic reconnection is closely associated with CMEs and solar flares. Magnetic reconnection is a phenomenon in which the sudden rearrangement of magnetic field lines occurs when two oppositely directed magnetic fields are brought together. Solar magnetic fields are stressed whose energy is released in the form of reconnection. These magnetic field lines can become twisted in a helical structure, with a 'right-hand twist' or a 'left hand twist'. CMEs acts as a 'valve' to release the magnetic energy being built up in the Sun's interior, as evidenced by the helical structure of CMEs, that would otherwise renew itself continuously each solar cycle and eventually rip the Sun apart. Hassler et al. (1999) pointed out that on the Sun, magnetic reconnection may happen on solar arcades—a series of closely occurring loops of magnetic lines of force. These lines of force quickly reconnect into a low arcade of loops, leaving a helix of magnetic field unconnected to the rest of the arcade. The sudden release of energy during this process causes the solar flare and ejects the CME. The helical magnetic field and the material that it contains may violently expand outwards forming a CME. This also explains why CMEs and solar flares typically erupt from what are known as the active regions on the Sun where magnetic fields are much stronger on average.

CME Impact on Earth

Earth directed CME ejection reaches it as an interplanetary CME (ICME) and the shock wave of traveling mass causes a geomagnetic storm deforming Earth's magnetosphere, compressing it on the day side and extending the night-side magnetic tail. When the magnetosphere reconnects on the night-side, it releases power on the order of terawatt scale, which is directed back toward Earth's upper atmosphere. Marsch et al.

(2005) investigated that solar energetic particles can cause particularly strong aurorae in large regions around Earth's magnetic poles. These are also known as the *Northern Lights* (aurora borealis) in the northern hemisphere, and the *Southern Lights* (aurora australis) in the southern hemisphere. Coronal mass ejections, along with solar flares of other origin, can disrupt radio transmissions and cause damage to satellites and electrical transmission line facilities, resulting in potentially massive and long-lasting power outages.Energetic protons released by a CME can cause an increase in the number of free electrons in the ionosphere, especially in the high-latitude polar regions. The increase in free electrons can enhance radio wave absorption, especially within the D-region of the ionosphere, leading to Polar Cap Absorption (PCA) events.Humans at high altitudes, as in airplanes or space stations, risk exposure to relatively intense solar particle events. The energy absorbed by astronauts is not reduced by a typical spacecraft shield design and, if any protection is provided, it would result from changes in the microscopic inhomogeneity of the energy absorption events. Coronal mass ejections are often associated with other forms of solar activity, most notably:

- Solar flares
- Eruptive prominence and X-ray sigmoids
- Coronal dimming (long-term brightness decrease on the solar surface)
- Moreton waves
- Coronal waves (bright fronts propagating from the location of the eruption)
- Post-eruptive arcades

The association of a CME with some of those phenomena is common but not fully understood. For example, CMEs and flares are normally closely related, but there was confusion about this point caused by the events originating beyond the limb. For such events no flare could be detected. Most weak flares do not have associated CMEs; most powerful ones do. Some CMEs occur without any flare-like manifestation, but these are the weaker and slower ones.It is now thought that CMEs and associated flares are caused by a common event (the CME peak acceleration and the flare impulsive phase generally coincide). In general, all of these events (including the CME) are thought to be the result of a large-scale restructuring of the magnetic field; the presence or absence of a CME during one of these restructures would reflect the coronal environment of the process (i.e., can the eruption be confined by overlying magnetic structure, or will it simply break through and enter the solar wind.

Theoretical Models for CME's

CMEs are initiated by the heat of an explosive flare as was hypothesized. Gonzalez et al. (1994) investigated that it became evident very fast that CMEs were not related with flares, and that even those that were often started before the flare. Because CMEs are initiated in the solar corona dominated by magnetic energy, their energy source must be magnetic. Because the energy of CMEs is so high, it is unlikely that their energy could be directly driven by emerging magnetic fields in the photosphere (although this is still a possibility). Therefore, most models of CMEs assume that the energy is stored up in the coronal magnetic field over a long period of time and then suddenly released by some instability or a loss of equilibrium in the field. There is still no consensus on which of these release mechanisms is correct, and observations are not currently able to constrain these models very well. These same considerations apply equally well to solar flares, but the observable signatures of these phenomena differ.

Interplanetary CME's

CMEs typically reach Earth one to five days after leaving the Sun. During their propagation, CMEs interact with the solar wind and the interplanetary magnetic field (IMF). Sugiura & Kamie (1991) pointed out that as a consequence, slow CMEs are accelerated toward the speed of the solar wind and fast CMEs are decelerated toward the speed of the solar wind. The strongest deceleration or acceleration occurs close to the Sun, but it can continue even beyond Earth orbit (1 AU), which was observed using measurements at Mars and by the Ulysses spacecraft. CMEs faster than about 500 km/s (310 mi/s) eventually drive a shock wave.This happens when the speed of the CME in the frame of reference moving with the solar wind is faster than the local fast magnetosonic speed. Such shocks have been observed directly by coronagraphs in the corona, and are related to type II radio bursts. They are thought to form sometimes as low as $2\,R_\odot$ (solar radii). They are also closely linked with the acceleration of solar energetic particles. The largest recorded geomagnetic perturbation, resulting presumably from a CME, coincided with the first-observed solar flare on 1 September 1859. The resulting solar storm of 1859 is now referred to as the Carrington Event. The flare and the associated sunspots were visible to the naked eye (both as the flare itself appearing on a projection of the Sun on a screen and as an aggregate brightening of the solar disc), and the flare was independently observed by English astronomers R. C. Carrington and R. Hodgson. The geomagnetic storm was observed with the recording magnetograph at Kew Gardens. The same instrument recorded a *crochet*, an instantaneous perturbation of Earth's ionosphere by ionizing soft X-rays. This could not easily be understood at the time because it predated the discovery of X-rays by Röntgen and the recognition of the ionosphere by Kennelly and Heaviside. The storm took down parts of the recently created US telegraph network, starting fires and shocking some telegraph operators.Historical records were collected and new observations recorded in annual summaries by the Astronomical Society of the Pacific between 1953 and 1960. The first detection of a CME as such was made on 14 December 1971, by R. Tousey (1973) of the Naval Research Laboratory using the seventh Orbiting Solar Observatory (OSO-7). The discovery image (256 × 256 pixels) was collected on a Secondary Electron Conduction (SEC) vidicon tube, transferred to the instrument computer after being digitized to 7 bits. Then it was compressed using a simple run-length encoding scheme and sent down to the ground at 200 bit/s. A full, uncompressed image would take 44 minutes to send down to the ground. The telemetry was sent to ground support equipment (GSE) which built up the image onto Polaroid print. David Roberts, an electronics technician working for NRL who had been responsible for the testing of the SEC-vidicon camera, was in charge of day-to-day operations. He thought that his camera had failed because certain areas of the image were much brighter than normal. But on the next image the bright area had moved away from the Sun and he immediately recognized this as being unusual and took it to his supervisor, Dr. Guenter Brueckner, and then to the solar physics branch head, Dr. Tousey. Earlier observations of *coronal transients* or even phenomena observed visually during solar eclipses are now understood as essentially the same thing.On 9 March 1989 a coronal mass ejection occurred. On 13 March 1989 a severe geomagnetic storm struck the Earth. It caused power failures in Quebec, Canada and short-wave radio interference.On 1 August 2010, during solar cycle 24, scientists at the Harvard–Smithsonian Center for Astrophysics (CfA) observed a series of four large CMEs emanating from the Earth-facing hemisphere of the Sun. The initial CME was generated by an eruption on 1 August that was associated with NOAA Active Region 1092, which was large enough to be seen without the aid of a solar telescope. The event produced significant aurorae on Earth three days later.on 23 July 2012, a massive, and potentially damaging, solar superstorm (solar flare, CME, solar EMP) occurred but missed Earth,an event that many scientists consider to be Carrington-class event.On 31 August 2012 a CME connected with Earth's magnetic environment, or magnetosphere, with a glancing blow causing aurora to appear on the night of 3 September.Geomagnetic storming reached the G2 (Kp=6) level on NOAA's Space Weather Prediction Center scale of geomagnetic disturbances.14 October 2014 ICME was photographed by the Sun-watching spacecraft PROBA2 (ESA), Solar and Heliospheric Observatory (ESA/NASA), and Solar Dynamics Observatory (NASA) as it left the Sun, and STEREO-A observed its effects directly at 1 AU. ESA's Venus Express gathered data. The CME reached Mars on 17 October and was observed by the Mars Express, MAVEN, Mars Odyssey, and Mars Science Laboratory missions. On 22 October, at 3.1 AU, it reached comet 67P/Churyumov–Gerasimenko, perfectly aligned with the Sun and Mars, and was observed by Rosetta. On 12 November, at 9.9 AU, it was observed by Cassini at Saturn. The New Horizons spacecraft was at 31.6 AU approaching Pluto when the CME passed three months after the initial eruption, and it may be detectable in the data. Voyager 2 has data that can be interpreted as the passing of the CME, 17 months after. The Curiosity rover's RAD instrument, *Mars Odyssey*, *Rosetta* and *Cassini* showed a sudden decrease in galactic cosmic rays (Forbush decrease) as the CME's protective bubble passed by.

5.9 F10.7cm flux

A unit used by astronomers to express the flux density of radio energy from the sun as received at the Earth. One solar flux unit = 10^{-22} watt per square meter-hertz. Symbol, sfu. 1 sfu = 10,000 jansky. The solar flux density is proportional to the sunspot number. Chapman & Ferraro (1930) pointed out that typically, the solar flux density is measured at a wavelength of 10.7 centimeters (approximately corresponding to a frequency of 2800 megahertz), because these measurements correlate with the sun's output in the ultraviolet. Values range from 67 sfu (when the sun shows no sunspots) to 300 sfu, but bursts can be much higher. The record is 55,000 sfu, recorded on *6 June 19*. Radio amateurs who delight in receiving distant stations (DX'ers) take great interest in sfu, because the long distance transmission of shortwave signals depends on reflecting signals off the ionosphere, and the state of the ionosphere depends on the solar flux. At 18 minutes past the hour, the NIST radio station WWV (Fort Collins, CO) broadcasts a measurement in sfu of the solar radio flux at 10.7 cm (2800 MHz), obtained from a radio telescope at the Penticton Radio Observatory in British Columbia. Readings are updated every three hours, beginning at midnight UTC. The intensity of radio emissions from the chromospheres and corona of the sun at a wavelength of 10.7cm (i.e., 2800 Mhz) has been found to be correlated well with the solar activity. Consequently, this measurement is used to quantify solar activity in place of the sunspot number. Flux measurements are given in units of solar flux : 1sfu = $10^{-22} Js^{-1} m^{-2} Hz^{-1}$. Data are tabulated as the observed F10.7cm flux and the adjusted F10.7cm flux. The former are the actual measured values and are affected by changing distance between the Earth and Sun throughout the year, while the later are scaled to a standard distance of 1AU. The observed F10.7cm flux values are useful in studies on ionosphere physics and other terrestrial effects of solar activity. The adjusted F10.7cm flux values are more descriptive of the Sun's activity. This measure of solar activity has clear advantage over the index like sunspot numbers and sunspot areas. This is basically due to the objective nature of F10.7cm flux value. This data can be measured on all weather conditions. From the year 1947 onwards, observations have been catalogued routinely by radio telescopes at the Alogonquin Radio Observatory(ARO), near Ottawa, Canada. Since 1991, Dominion Radio Astrophysical Observatory, near Penticton, British Columbia is measuring the F10.7cm solar flux data. Measurements of this flux have been taken daily by the canadian solar radio monitoring programme since 1946. The relationship between the F10.7cm radio flux and international sunspot number(R) has been established through their correlation. In this model two fact have been accounted. They are, first of all the F10.7cm radio flux has a base level of about 67 solar flux units. Secondly, the slope of the relationship changes as the sunspot number increases up to about 30. This is modelled in as Equation (1.6) [27,32] as: F10.7 = 67 + 0.97R + 17.6(exp(−0.035R)). (1.6)

5.10 *K*-index

The ***K*-index** quantifies disturbances in the horizontal component of earth's magnetic field with an integer in the range 0–9 with 1 being calm and 5 or more indicating a geomagnetic storm. It is derived from the maximum fluctuations of horizontal components observed on a magnetometer during a three-hour interval. The label *K* comes from the German word *Kennziffer* meaning *"characteristic digit"*. The ***K*-index** was introduced by Julius Bartels in 1939.

Calculation of *K*-index

The *K*-scale is quasi-logarithmic. The conversion table from maximum fluctuation *R* (in units of nanoteslas, nT) to *K*-index, varies from observatory to observatory in such a way that the historical rate of occurrence of certain levels of *K* are about the same at all observatories. In practice this means that observatories at higher geomagnetic latitude require higher levels of fluctuation for a given *K*-index. For example, at Godhavn, Greenland, a value of *K* = 9 is derived with *R* = 1500 nT, while in Honolulu, Hawaii, a fluctuation of only 300 nT is recorded as *K* = 9. In Kiel, Germany, *K* = 9 corresponds to *R* = 500 nT or greater. The real-time *K*-index is determined after the end of prescribed intervals of 3 hours each: 00:00–03:00, 03:00–06:00, ..., 21:00–24:00. The maximum positive and negative deviations during the 3 hour period are added together to determine the total maximum fluctuation. These maximum deviations may occur any time during the 3 hour period.

The K_p-index and estimated K_p-index

The official planetary K_p-**index** is derived by calculating a weighted average of *K*-indices from a network of geomagnetic observatories. Since these observatories do not report their data in real-time, various operations centers around the globe estimate the index based on data available from their local network of observatories. The K_p-index was introduced by Bartels in 1939.

The relationship between K and A

The A-index provides a daily average level for geomagnetic activity. Because of the non-linear relationship of the K-scale to magnetometer fluctuations, it is not meaningful to take the average of a set of K-indices. The K_p-index is used for the study and prediction of ionospheric propagation of high frequency radio signals. Geomagnetic storms, indicated by a $K_p = 5$ or higher, have no direct effect on propagation. However they disturb the F-layer of the ionosphere, especially at middle and high geographical latitudes, causing a so-called *ionospheric storm* which degrades radio propagation. The degradation mainly consists of a reduction of the maximum usable frequency (MUF) by as much as 50%.Sometimes the E-layer may be affected as well. In contrast with sudden ionospheric disturbances (SID), which affect high frequency radio paths near the Equator, the effects of ionospheric storms are more intense in the polar regions. The **K-index** is used to ratify disturbances in the horizontal component of earth's magnetic field by numbering them by integers from 0–9.Integer 1 represents calm and 5 or higher values indicates a geomagnetic storm. It is acquired from the maximum fluctuations of horizontal components observed on a magnetometer during a three-hour interval.

5.11 Results

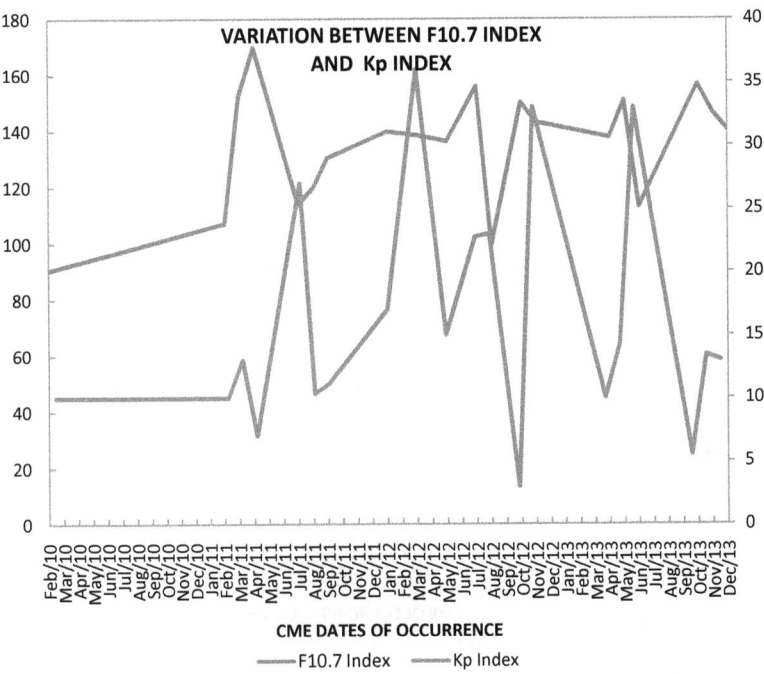

Fig 5.1 Variation between kp index and f10.7 index on 51 cme dates of occurrence during solar cycle 24

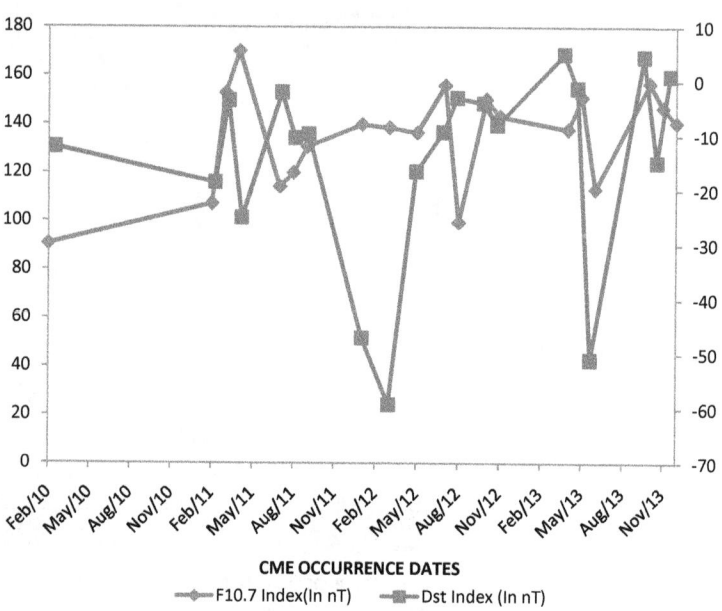

Fig 5.2 Variation between f10.7 index and dst index on 51 cme dates of occurrence during solar cycle 24

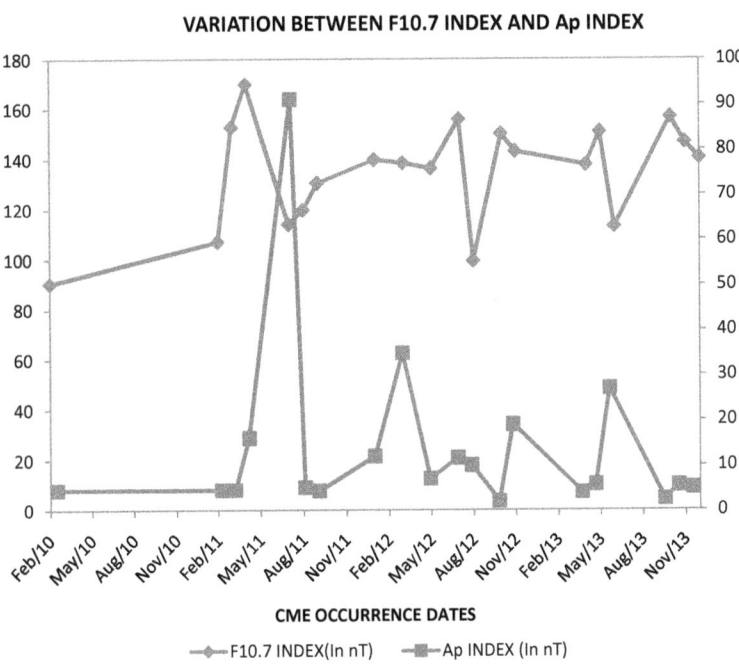

Fig 5.3 Variation between f10.7 index and ap index on 51 cme dates of occurrence during solar cycle 24

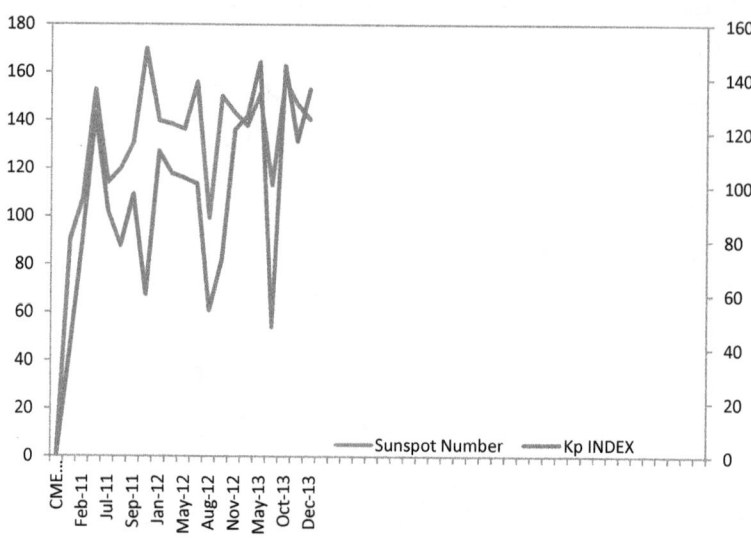

Fig. 5.4 Variation between sunspot number and kp index on 51 CME dates of occurrence during solar cycle 24

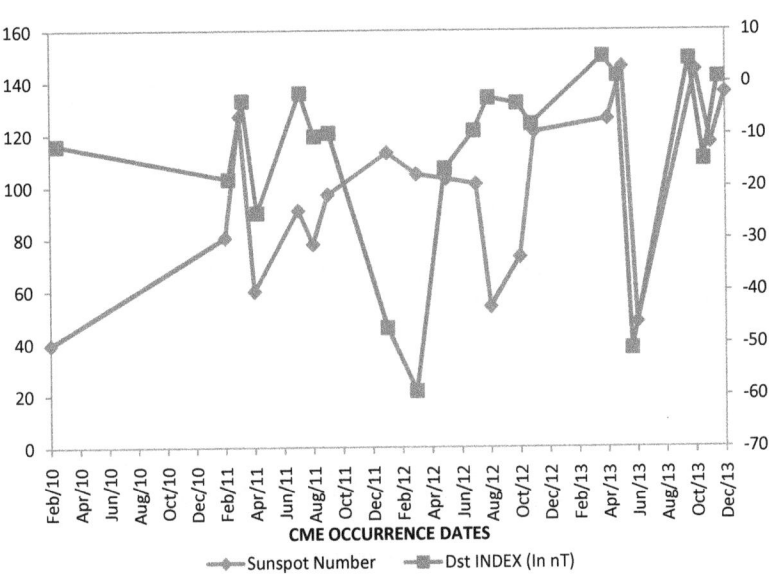

Fig. 5.5 Variation between sunspot number and dst index on 51 CME dates of occurrence during solar cycle 24

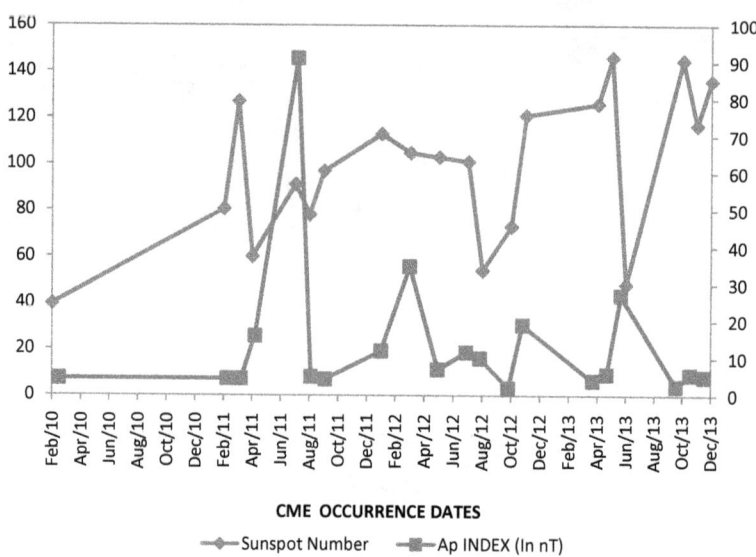

Fig. 5.6 Variation between sunspot number and ap index on 51 CME dates of occurrence during solar cycle 24

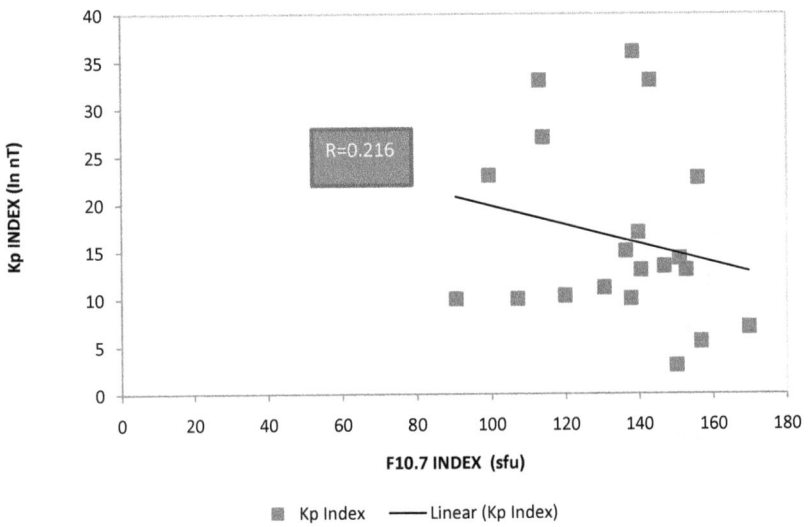

Fig. 5.7 Scatter plots between f10.7 index and kp index on 51 CME dates of occurrence during solar cycle 24

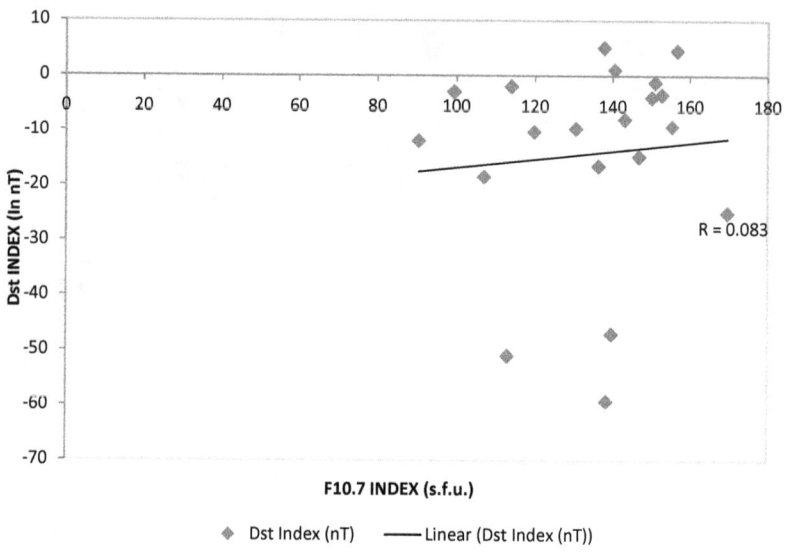

Fig 5.8 Scatter plots between f10.7 index and dst index on 51 CME dates of occurrence during solar cycle 24

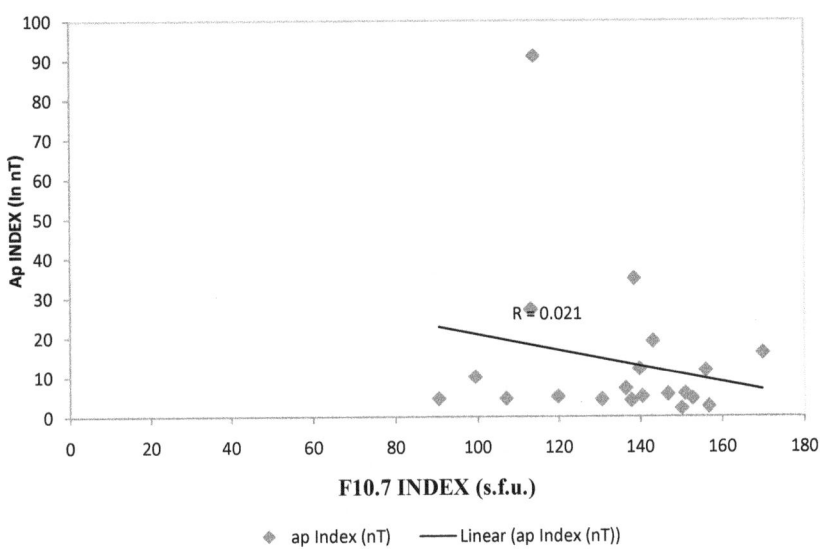

Fig 5.9 Scatter plots between f10.7 index and ap index on 51 CME dates of occurrence during solar cycle 24

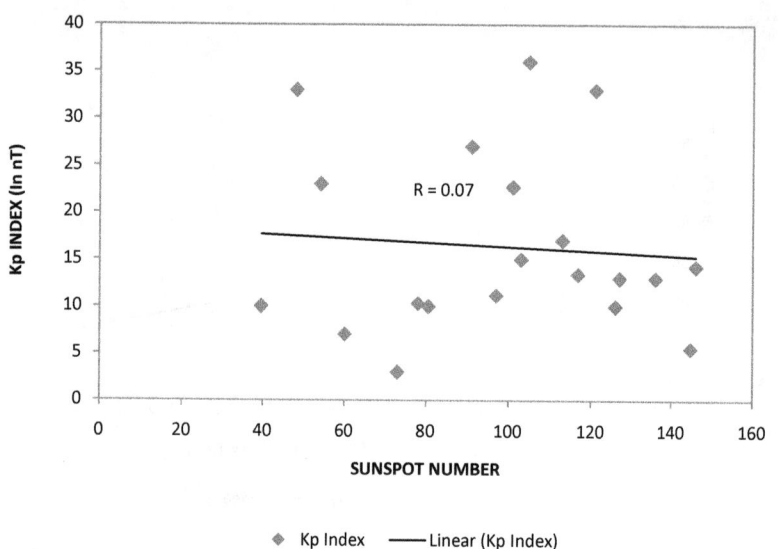

Fig 5.10 Scatter plots between sunspot number and kp index on 51 CME dates of occurrence during solar cycle 24

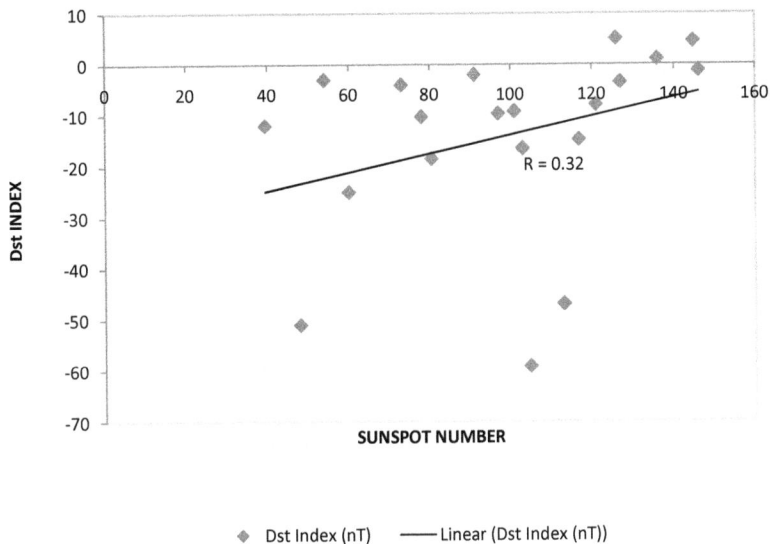

Fig 5.11 Scatter plots between sunspot number and dst index on 51 cme dates of occurrence during solar cycle 24

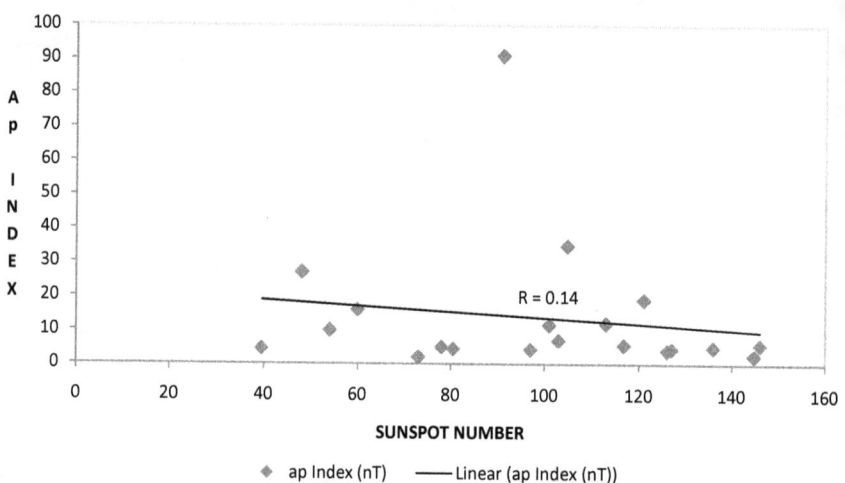

Fig 5.12 Scatter plots between sunspot number and ap index on 51 CME dates of occurrence during solar cycle 24

5.12 Discussion

Solar parameters and earth's atmospheric parameters during the 51 CME's and SSN or F10.7-cm of solar radio flux have been analyzed, solar activity parameters which have been related with earth's atmospheric parameters such as Dst index, Kp index and Ap index are also in cooperated in this study. Time Series and Scatter Plots analysis have been done. Time Series variation between F10.7-cm. solar radio flux and Kp index and partial linear and partial non-linear relationship between the two variables Both F10.7-cm. solar radio flux and Kp index increases abruptly to their maximum value in 2011 when solar activity starts increasing in the ascending phase. They decrease abruptly after reaching to their maximum values. Both showed a linear relationship with an exception in 2013 since increase in F10.7-cm. solar radio flux disturbs the horizontal component of earth's magnetic field. Alongwith it , some prominent radio blackouts occurs in upper earth atmosphere causing X-ray bursts and leading to an increase in F10.7-cm. solar radio flux and a decrease in Kp index.Their maximum values are 170 sfu and 21 nT respectively. F10.7-cm. solar radio flux and Dst index starts increasing in 2011 when solar activity starts series of CME's which were emitted during 2011 ,2012 and 2013 .Dst index reaches abruptly to its lowest values during 2012 and 2013 due to sudden decrease in CRI during that period.Highest and lowest values of Dst index are 5 nT and -60 nT respectively and of F10.7-cm. solar radio flux are 170 sfu and 90 sfu respectively. F10.7-cm. solar radio flux and Ap index remains constant from 2008 to 2011.In 2011, when solar activity increases to its maximum value, F10.7-cm. solar radio flux and Ap index reaches to their maximum values and suddenly drops. Majority of values(90 %) of both parameters shows a linear relationship whereas other values shows a non-linear relationship.Highest and lowest values of F10.7-cm. solar radio flux and Ap index are 170 sfu and 95 nT respectively and lowest values are 90 sfu and 8 nT respectively.Both exhibited a non-linear relationship during 2011 due to many prominent X-ray bursts. correlation coefficient "($R = 0.07$)" was observed. This was due to solar minima continuing throughout the entire period of observation due to which sunspot number was predominantly low. This led Kp index towards an abnormal variation. Scatter plots between Sunspot number and Dst index exhibits a very strong correlation ($R = 0.32$). An upward moving trendline with tighter data points falling along the trendline was observed

.Reason behind this strong correlation has been a constant unexpectedly low solar activity during the period of observation.This led to a constant sunspot number value neither low nor high ,hence the Dst index values also varied equally with sunspot number variations. Using data from Omni web (Website: https://omniweb.gsfc.nasa.gov/html/ow_data.html) which is a website of Goddard Space Flight Centre ,NASA ,there is a weak positive correlation between F10.7 index and Kp index .A very small value of R = 0.216 indicates an increase in Kp index values with increase in F10.7 index values but this variation is not pronounced as the data points are not tighter along the trendline.It is also observed in the Time Series Analysis between the two parameters.This weak correlation is due to the fact that many flares arose during this period ,predominantly earth- directed .Sunspot number also exhibited an exponential rise and fall during this time .Both parameters have not showed a strict linear correlation . A feeble value of R=0.1 indicates non-linear relationship between the two variables.The time series graph also advocated the fact that change in F10.7 index have not rigorously led to change in Dst index values.The monster sunspot erupted releasing a very predominant earth-directed solar flare into space .Negative Dst values were prevalent during this period, due to which earth's magnetic field got weakened. A very small value of R = 0.021 points towards a weak positive correlation between the two variables .The period of observation of the analysis has witnessed merely 40% of the geomagnetic activities, leading to this weak correlation.

Both SSN and Kp index varies linearly in the ascending phase of solar cycle 24 ,with an exception during the first quarter of 2011.Highest and lowest values of SSN are 170 and 110 respectively and of Kp index are 165 nT and 60 nT respectively.SSN achieves its highest value in the second quarter of 2011 and Kp index achieves its highest value in the second quarter of 2013. A very poor value of correlation coefficient "(R = 0.14)" indicates a very weak positive correlation between sunspot number and Ap index .Since Ap index indicates the average value of geomagnetic storms ,low value of correlation coefficient shows abnormal change in the magnitude of geomagnetic storm values with increase in sunspot number values. SSN and Dst index linearly in the ascending phase of Solar cycle 24.SSN achieves its highest value of 142 in the mid of June 2013 and achieves a lowest value of 40 in the beginning of 2010.Dst index achieves a highest value of 5 nT in the first quarter of 2013 and a lowest value of -60 nT in the first quarter of

2012. Both parameters exhibited a linear relationship because SSN is directly proportional the number of solar protons and electrons forming the ring current SSN and Ap index exhibits partial linear and partial non-linear variation during the ascending phase of Solar cycle 24. SSN achieved a highest value of 145 in the first quarter of 2013 and a lowest value of 40 in the beginning of 2010 and Ap index achieved a highest value of 150 nT in the mid of 2011 and a lowest value of 2 nT in the last quarter of 2013. Dst index exhibits a simultaneous time variation with Sunspot number and F10.7 index. Dst index is the main parameter of earth's atmospheric weather which exhibits a strong positive correlation with Sunspot Number .Increase in Solar Flux Units (F10.7 Index) can lead to a very high magnitude of Geomagnetic storm ,having a magnitude of G4 which is a very severe level magnitude ,leading to very high power outages during solar cycle 24. This trend has been the reverse as observed in Solar cycle 23 where increase in F10.7 index led to a decrease in Ap index and other earth's atmospheric parameters. A modeling between Sunspot number and Dst index will definitely go a long way in forecasting space weather and will help to mitigate the harmful effects of space weather on earth like damage to satellites, power grid failures, climate changes, etc.

CHAPTER 6
WAVE ANALYSIS OF COSMIC RAYS ASSOCIATED WITH SOLAR WIND AND GEOMAGNETIC

6.1 Introduction

On observing the wavelet spectrum density (WSD) of CR for the epoch 1953-2006 of different neutron monitor stations like Climax , it can be inferred that the time evolution of the ~ 1.3 year wave in CR to the signal was obtained also reported earlier in interplanetary parameters and in geomagnetic activity e.g. by Mursula and Zieger as well as by Obridko and Shelting in solar magnetic fields since 1915 inferred from H-alpha filament observations. Recurrence in CR intensity at ~ 1.7 years was reported, analyzed by wavelet methods found also in outer heliosphere and certified by data of Voyager. The 20 months power spectrum shape is changing as has been enunciated using neutron monitor (NM) data of Calgary and Deep River neutron monitor stations. The wavelet spectrum analysis shows that 1.3 year wave density profile is different from that of the 1.7 year. In sunspot groups and flare index the periodicities in that range have shown differences in the solar hemispheres. Recently Vecchio et al performed the detail analysis of different components of heliomagnetic field for 1976-2003. The authors found that quasi-biennial oscillations (QBO) are also identified as a fundamental timescale of variability of the magnetic field and associated with poleward magnetic flux migration from low latitude around the maximum and descending phase of solar activity cycle. The quasi periodicity in CR shorter than ~11 years have been reported by using different methods from data 1953-1996 e.g. by Mavromichalaki et al . Periodicities ~11 yr and ~22 yr along with their origin are discussed e.g. by Venkatesan and Badruddin . Lomb-Scargle Periodogram of Climax NM data (1953 – 2006) indicates several quasi periodicity at very low frequencies (~5.5, ~6.4, ~ 8.2 yr). Since nowadays there exist rather long time series of CR measured directly both on the ground as well as in stratosphere at various depths, it is worth to check the trend of three-solar cycle recurrency reported for the first time by Ahluwalia from the ground based CR measurements. The analysis done by Pérez-Peraza et al. using the cosmogenic ^{10}Be nuclide as an indirect proxy of CR over the long time period, has shown common frequency of 30± 2 years which appears also in the time .Longer time series of CR, solar activity and IMF were examined recently. Here we use the direct stratospheric measurements of CR provided routinely over long time period as well as data. The periodograms in Figure 1 indicate the tendency that the three-cycle trend is present in both data sets.Discussion about ~ 27 d periodicity in ground based CR recordings continues in accordance with obtaining longer time series of NMs and muon telescope (MT) data. Recently this has been discussed in detail e.g. in papers. Agarwal indicates that ~27 d CR variation correlates with B, Bz, v, and B(v x B). Wavelet method provides fine structure of appearing q-pers in the signal as inspected over long time. For example the profile of WSD using Climax NM data over 1953-2006 indicates the double structure during year 1986, with two peaks, one around ~27 days and another ~30-31 days. The two peak structure is similar to that reported by Dunzlaff et al for GCR from measurements at lower energies, EPHIN on SOHO. Gil and Alania reported the 3–4 cycling structure of ~ 27 day quasi periodic amplitude in NM data with cut-off rigidities below 8 GV. Analysis of data from Nagoya MT (Sabbah, personnal communication) were checked by periodogram technique and we found that the ~3 Carrington rotation q-per is significant even at higher energies of primaries . Modzelewska and Alania discussed the 3D model of ~ 27 day CR variations.

6.2 Solar Wind

This solar plasma mostly consists of electrons, protons and alpha particles constituting the cosmic rays with kinetic energy between 0.5 and 10 keV. Embedded within the solar-wind plasma is the interplanetary magnetic field. The solar wind varies in density, temperature and speed over time and over solar latitude and longitude. Solar corona has a high energy due to its high temperature resulting from coronal magnetic field , this inturn makes solar wind particles escape out from corona. A current of charged particles released from the topmost atmosphere of the Sun, called the corona is the solar wind.

6.3 Cosmic Rays Associated with Solar Wind

At a distance of more than a few solar radii from the Sun, the solar wind reaches speeds of 250 to 750 kilometers per second and becomes supersonic, reaching speed faster than the speed of the magnetosonic wave. Heliosphere the huge, bubble-like cavity formed and encompassed by the Sun is also affected by Solar Wind originating from the Sun. Solar wind flows unobstructed along the Solar System for large distances until it comes across the termination shock, where its motion slows abruptly due to the outside pressure of the interstellar medium. The existence of solar wind cosmic particles flowing outward from the Sun to the Earth was first suggested by British astronomer Richard C. Carrington. In 1859, Carrington alongwith Richard Hodgson made the first examination of solar wind ,which was renowned later as a solar flare. This is a sudden, restricted increase in brightness on the solar disc, now known to often occur in conjunction with an episodic ejection of material and magnetic flux from the Sun's atmosphere, known as a coronal mass ejection. Geomagnetic storms were observed in subsequent days, and Carrington suspected that there might be a connection, attributed to the arrival of the coronal mass ejection in near-Earth space and its subsequent interaction with the Earth's magnetosphere. George FitzGerald later suggested that matter was being regularly accelerated away from the Sun and was reaching the Earth after several days.

In 1910 British astrophysicist Arthur Eddington essentially suggested the existence of the solar wind. Kristian Birkeland. enundated that the ejected material consisted of both ions and electrons. His geomagnetic surveys showed that auroral activity was nearly uninterrupted. As these displays and other geomagnetic activity were being produced by particles from the Sun, he concluded that the Earth was being continually bombarded by "rays of electric corpuscles emitted by the Sun". In 1916, Birkeland proposed that solar rays are inclusively negative and positive rays. In 1919, Frederick Lindemann also suggested that particles of both polarities, protons as well as electrons, come from the Sun. These phenomenons propounded by Carrington ,Hodgson ,Birkeland and Lindemann proved that primarily cosmic rays are associated with the Solar Wind .

Until 1960, it was clear that thermal acceleration alone cannot account for the high speed of solar wind. An additional unknown acceleration mechanism is required and likely relates to magnetic fields in the solar atmosphere.The Cosmic rays belonging to the Solar Wind , emerges out from the Sun's corona. Maxwellian distribution explains the range and distribution of speeds of particles within the internal corona .Resultantly due to thermal collisions,the mean velocity of these particles is about 145 km/s, which is well below the solar escape velocity of 618 km/s.It has been investigated however, that a few of the particles achieve energies sufficient to reach the terminal velocity of 400 km/s, which allows them to feed the solar wind. At the same temperature, electrons, due to their much smaller mass, reach escape velocity and build up an electric field that further accelerates ions away from the Sun. The total number of Cosmic rays carried away from the Sun by the solar wind is about 1.3×10^{36} per second. Thus, the total mass loss each year is about $(2-3) \times 10^{-14}$ masses, or about 1.3–1.9 million tonnes per second. This is equivalent to losing a mass equal to the Earth every 150 million years. However, only about 0.01% of the Sun's total mass has been lost through the solar wind. Other stars have much stronger stellar winds that result in significantly higher mass loss rates.The cosmic rays arising from the solar wind in two fundamental states, the slow solar wind and the fast solar wind. The slow solar wind is observed to have a velocity of 300–500 km/s, a temperature of $\sim 10^5$ K and a configuration similar to the corona. The fast solar wind has a specific velocity of 750 km/s, a temperature of 8×10^5 K and configuration similar to the Sun's photosphere. The slow solar wind is twice as dense as and more variable in nature than the fast solar wind.

The slow Cosmic rays originates from a region around the Sun's equatorial belt known as the "streamer belt", where coronal streamers are produced by magnetic flux open to the heliosphere draping over closed magnetic loops. Observations of the Sun between 1996 and 2001 showed that emission of the slow solar wind occurred at latitudes up to 30–35° during the solar minimum (the period of lowest solar activity), then expanded toward the poles as the solar cycle approached maximum.The fast Cosmic rays originates from coronal holes, which are funnel-like regions of open field lines in the Sun's magnetic field.Such open lines are particularly prevalent around the Sun's magnetic poles. The plasma source is small magnetic fields created by convection cells in the solar atmosphere. These fields confine the plasma and transport it into the narrow necks of the coronal funnels, which are located only 20,000 kilometers above the photosphere. The plasma is released into the funnel when these magnetic field lines reconnect.

6.4 Magnetospheres

Where the Cosmic rays developed by the solar wind intersects with a planet that has a well-developed **magnetic field** (such as Earth, Jupiter or Saturn), the particles are deflected by the **Lorentz force**. This region, known as the **magnetosphere**, causes the particles to travel around the planet rather than bombarding the atmosphere or surface. The magnetosphere is roughly shaped like a **hemisphere** on the side facing the Sun, then is drawn out in a long wake on the opposite side. The boundary of this region is called the **magnetopause**, and some of the particles are able to penetrate the magnetosphere through this region by partial reconnection of the magnetic field lines.

6.5 Earth Atmospheres

The Cosmic rays associated with a solar wind interacts with planetary atmospheres. Planets with a weak or non-existent magnetosphere are subject to atmospheric stripping .Earth is largely protected from the Cosmic rays emanating from the solar wind by its magnetic field, which deflects most of the charged particles; however some of the charged particles are trapped in the Van Allen radiation belt , which is a region of energetic charged particles, mostly originating from the solar wind .Earth has two such belts and sometimes others are temporarily created. The belts are named after their discoverer James Van Allen.Earth's two main belts extend from an altitude of about 640 to 58,000 km (400 to 36,040 mi) above the surface in which region radiation levels vary. Most of the particles that form the belts are thought to come from solar wind and other particles by cosmic rays.By trapping the solar wind, the magnetic field deflects those energetic particles and protects the atmosphere from destruction.

6.6 Geomagnetic Storm

A **geomagnetic storm** (commonly referred to as a **solar storm**) is a temporary disturbance of the Earth's magnetosphere caused by a shock wave associated with

Cosmic rays of solar wind and/or cloud of magnetic field emanated from solar wind Cosmic Rays that interacts with the Earth's magnetic field.

Cosmic rays associated with a Geomagnetic Storm

The Geomagnetic storms are quantified by Geomagnetic Index or *K*-**index** which evaluates disturbances in the horizontal component of earth's magnetic field with an integer in the range 0–9 with 1 being calm and 5 or more indicating a geomagnetic storm. It is derived from the maximum fluctuations of horizontal components observed on a magnetometer during a three-hour interval. The label *K* stands for *Kennziffer* which is a German word meaning "*characteristic digit*". The Cosmic rays formed by solar coronal mass ejection (CME) or a co-rotating interaction region (CIR) ultimately forms geomagnetic storm which is a high-speed stream of solar wind originating from a coronal hole.Its frequency increases and decreases with the sunspot cycle. During solar maximum, geomagnetic storms formed due to the Cosmic rays of CME or CIR occurs more often, with the majority driven by CMEs. During solar minimum, storms are mainly driven by CIRs (though CIR storms are more frequent at solar maximum than at minimum).Enhancement in the pressure of Cosmic rays compresses the magnetosphere. The Cosmic ray magnetic field interacts with the Earth's magnetic field and transfers an increased energy into the magnetosphere. Both interactions cause an increase in plasma movement through the magnetosphere (driven by increased electric fields inside the magnetosphere) and an increase in electric current in the magnetosphere and ionosphere. During the main phase of a geomagnetic storm, electric current in the magnetosphere creates a magnetic force that pushes out the boundary between the magnetosphere and the solar wind. In 1931, Sydney Chapman and Vincenzo C. A. Ferraro wrote an article, *A New Theory of Magnetic Storms*, explaining the phenomenon of formation of geomagnetic storms by solar wind Cosmic rays. They propounded that when Sun emits a solar flare it also emits a plasma cloud, now known as a coronal mass ejection ,which is a signature of cosmic rays of solar wind. They wrote that the cloud then compresses the Earth's magnetic field and thus increases this field at the Earth's surface.

The solar wind forms cosmic rays which carries with it the Sun's magnetic field. This field will have either a North or South orientation. If the solar wind has energetic bursts, it contracts and expands the magnetosphere, or if the solar wind takes a southward polarization, geomagnetic storms can be expected. The southward field causes magnetic reconnection of the dayside magnetopause, rapidly injecting magnetic and particle energy into the Earth's magnetosphere.

Geomagnetic Storm Effects

It has been suggested that a geomagnetic storm on the scale of the solar storm of 1859 today would cause billions or even trillions of dollars of damage to satellites, power grids and radio communications, and could cause electrical blackouts on a massive scale that might not be repaired for weeks, months, or even years.Such sudden electrical blackouts may threaten food production. Geologists and geodetic surveyors explore oil, gas and mineral deposits when the Earth's magnetic field is quiet ,i.e. when lesser amount of cosmic rays intercepted by solar wind interact with the earth's surface so that true magnetic signatures can be detected.Some geophysicists prioritize to work during geomagnetic storms, when strong variations in the Earth's normal subsurface electric currents allow them to sense subsurface oil or mineral structures.For these reasons, many surveyors use geomagnetic alerts and predictions to schedule their mapping activities.

6.7 Methodology

Wave Analysis of Cosmic Ray Intensity

In this study we have studied the wave analysis of amplitude and phase of the 21 Low Amplitude Cosmic ray events reaching earth .Amplitude wave is a strictly triangular wave with an exception in the beginning with an abrupt fall ,signifying a very low sunspot number in the beginning of year 2011 ,accompanied by constant increase in sunspot numbers with a few exceptions. Phase wave is a plane wave showing a steep rise and fall. The Cosmic Ray Intensity values of 3 major Low Amplitude events have been taken from 2011 to 2017 and the Fast Fourier Transform have been plotted.The Cosmic ray intensity data have been taken from the Bartol Research Institute's Neutron Monitor Station situated in the University of Delaware.(Website: neutron.bartol.udel.edu) For High Amplitude Events and Low Amplitude Events , GCR intensity displays a uniform variation with Solar wind velocity. There exists a positive association between SSN and Dst index. Ap index varies linearly with CRI for Low Apmplitude events of CRI during the entire solar cycle 24. Low Amplitude Events are weakly dependent on the Solar Wind Velocity.Amplitude and Phase Waves follows the same trend , i.e. CRI varies non-linearly with phase of solar cycle , with 90% of magnitude of amplitude and phase lying in the 0.2-0.5 hertz frequency range.Whenever CRI is greater ,it leads to occurrences of geomagnetic storms. A modeling between CRI and Ap index will go a long way to predict occurrences of geomagnetic storms.

Fig 6.1 Movement of Cosmic rays influenced by the Heliospheric Current Sheet
https://en.wikipedia.org/wiki/Heliospheric_current_sheet

6.8 Results

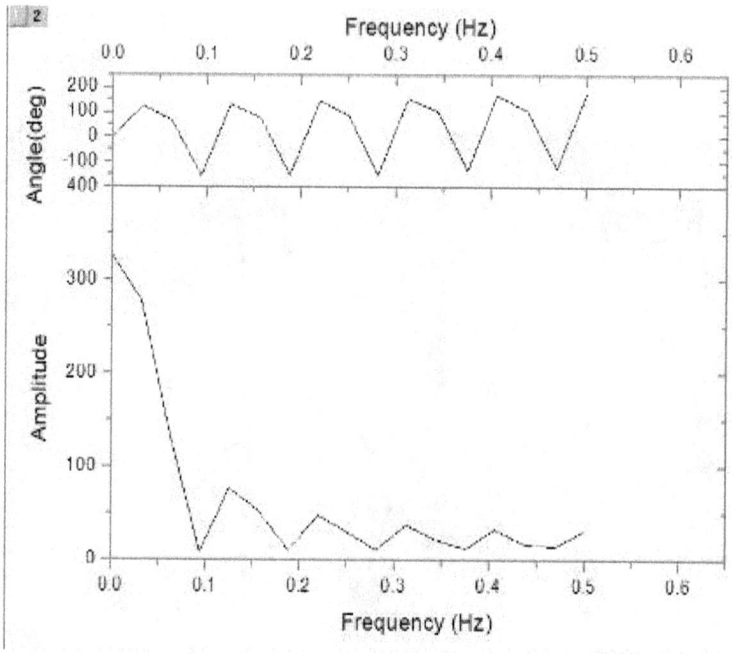

Fig 6.2 Amplitude Graph of Fourier Analysis of Cosmic Ray Intensity

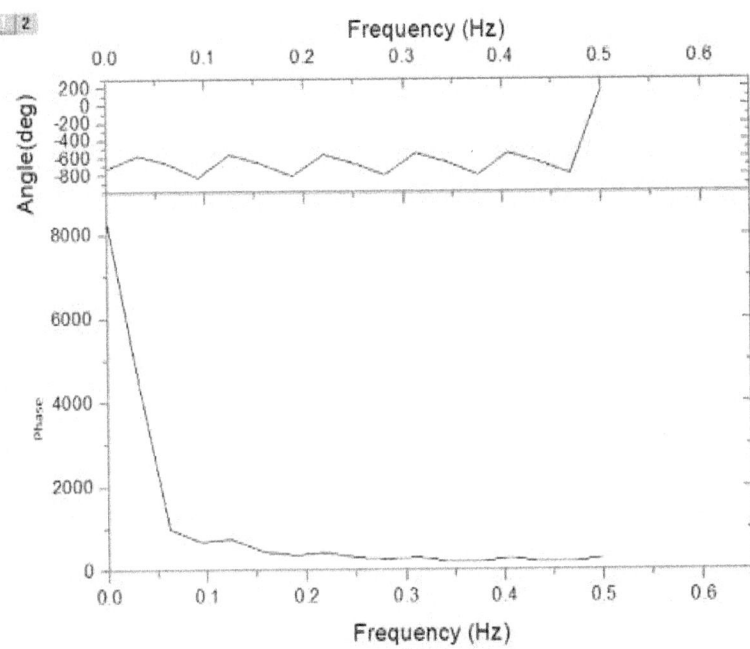

Fig 6.3 Phase Graph of Fourier Analysis of Cosmic Ray Intensity

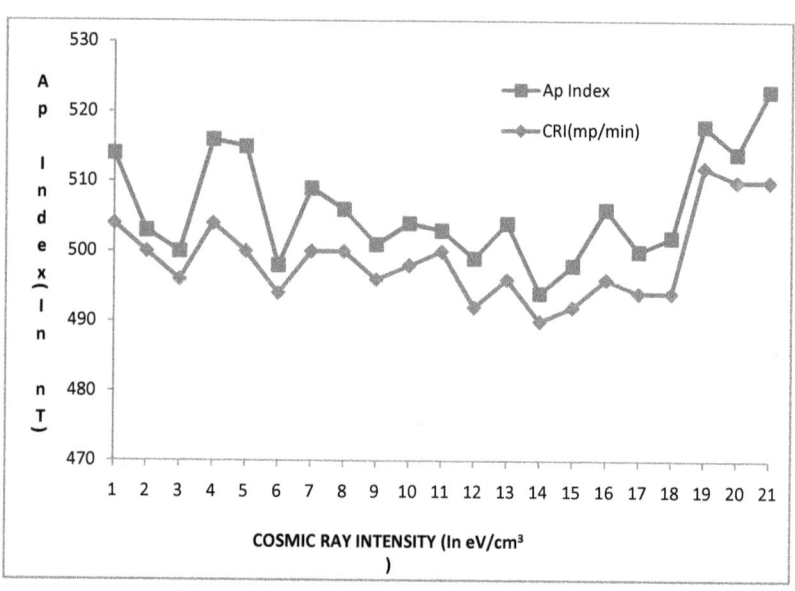

Fig 6.4 Graph Showing Variation between Ap Index and Cosmic Ray Intensit Above plot shows the variation in Ap index with cosmic ray intensity

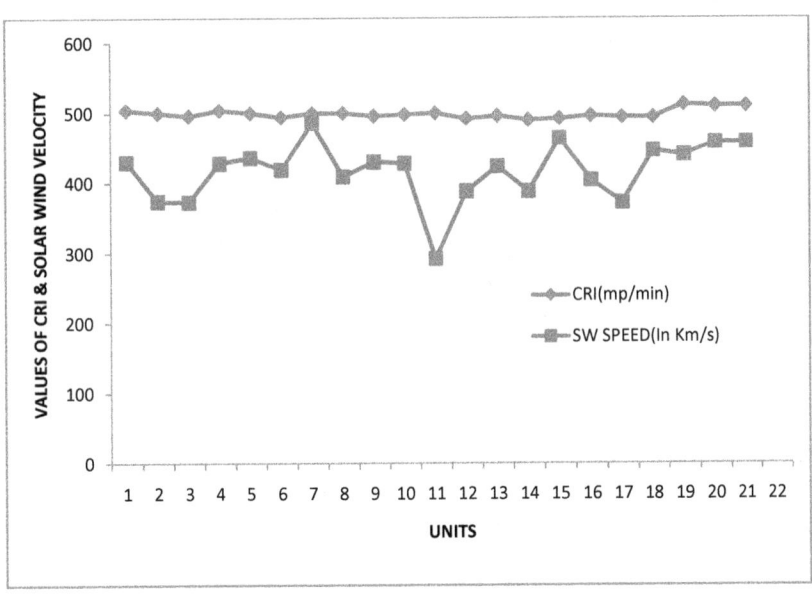

Fig 6.5 Graph showing Variation in Cosmic Ray Intensity with Solar Wind Velocity Above plot shows the variation in Solar Wind Velocity with Cosmic ray intensity

6.6 Conclusions

The wave analysis of amplitude and phase of the 21 Low Amplitude Cosmic ray events reaching earth .Amplitude wave is a strictly triangular wave with an exception in the beginning with an abrupt fall ,signifying a very low sunspot number in the beginning of year 2011 ,accompanied by constant increase in sunspot numbers with a few exceptions. Phase wave is a plane wave showing a steep rise and fall .It helps in understanding space weather.Low amplitude events are less or weakly dependent on the solar wind velocity.The geomagnetic activity index Ap on an average basis remains low during the period of each HAE/LAE with an exception in the year 2017 ,when it reached a high value of 526 ,due to a sudden increase in the number of sunspots compared to the time when there were no sunspots. This is because the interplanetary disturbances responsible for cosmic-ray modulation effects have not reached the Earth till that period of abrupt increase since prior to that , there were no sunspots. Further, the solar activity on the back side is not likely to produce the usual terrestrial manifestations such as geomagnetic storms, produced by activity on the visible side of the Sun.

CHAPTER 7
EFFECT OF SOLAR OUTPUT ON HIGH AND LOW AMPLITUDE WAVE TRAINS OF DIURNAL VARIATION OF COSMIC RAY INTENSITY

7.1 Introduction

Diurnal and semi-diurnal components helps to analyse basically the Cosmic ray anisotropic variations and their characteristics and the extent of the isotropic intensity collectively provides fingerprint for identifying the modulating process and the electromagnetic state of interplanetary space in the vicinity of the Earth. Yearly average values of the first harmonic of solar daily variation experience strong changes from year to year and with the cycle of solar activity. Amplitude and phase of diurnal anisotropy is changed with the solar activity cycles. The solar diurnal and semi-diurnal variations have been studied for many years. There are three prominent gradients of cosmic rays ,the radial gradient, the parallel mean-free path and the symmetric latitude gradient .The techniques propounded by Bieber and Chen has been used to derive the radial gradient, parallel mean-free path and symmetric latitude gradient of cosmic rays for rigidities < 200 GV. The radial gradient varies with the 11-year solar activity cycle whereas the parallel mean-free path appears to vary with the 22-year solar magnetic cycle. The symmetric latitudinal gradient reverses at each solar polarity reversal. Nagashima, Fujimoto and Jacklyn proposed a narrow Tail-In source anisotropy and separate Loss-Cone anisotropy as being responsible for the observed variations.They found that the Tail-In anisotropy is asymmetric and that both anisotropies had different positions from the prediction. Most 27-day modulations are observed at neutron monitor rigidities but not so readily at higher rigidities. An exception to this is the Isotropic Intensity Wave modulation observed in the early 1980s and again in 1991. This modulation is very strongly related to the heliospheric sector structure and implies a significantly different cosmic ray density on either side of the neutral sheet. The interpretation of most cosmic ray modulation phenomena requires good latitude coverage in both hemispheres.A close relationship was found between the magnitude and frequency of Forbush decreases and the eleven-year cosmic ray variation which deduced that the effect of Forbush and other transient decreases is a dominant factor in the long-term intensity modulation. It was proved that annual means of the CR diurnal anisotropy resulted from the addition of two distinct components. One, W has its maximum in the asymptotic direction of 128° E of the Sun and is well approximated by a wave W with a period of two solar cycles and the other component V has its maximum in the asymptotic direction 90° E of the Sun. It has been outlined that diurnal anisotropy is unidirectional during 1957-70 with direction along 1800 Hr LT (East-West) and during 1971-79 it consists of two components; one is in the East-West direction and the other is the radial component with direction along 1200 Hr LT. The direction of the dominant anisotropy vector points towards the 1800 Hr LT direction during the negative state of the solar cycle and toward earlier hours during the positive state. Kp and Ap has been correlated with the mean fluctuations in amplitude of IMF, which in turn is related to diffusive component of convection-diffusion theory. Ap is also found to related with solar wind velocity, which is related to the convective component of convection-diffusion theory. Daily variation during days of low and high amplitude anisotropic wave trains have been studied.The main factors acting as solar output are solar irradiance and solar variation impacting the High Amplitude Wave Trains of Diurnal Variation of Cosmic Ray Intensity .Solar irradiance and solar variation has been a main driver of climate change over the millennia to giga years of the geologic time scale. The power per unit area received from the Sun in the form of electromagnetic radiation is **Solar irradiance**. It is often consolidated over a given time period to report the radiant energy emitted into the surrounding environment, during that time period. Irradiance may be measured in space or at the Earth's surface after atmospheric absorption and scattering. Irradiance in space is a function of distance from the Sun, the solar cycle, and cross-cycle changes. Irradiance on the Earth's surface additionally depends on the tilt of the measuring surface, the height of the sun above the horizon, and atmospheric conditions. Solar irradiance affects plant metabolism and animal behavior.

The **solar cycle** is a nearly periodic 11-year change in the Sun's activity measured in terms of variations in the number of observed sunspots on the solar surface. The sunspot time series is the longest, continuously observed (recorded) time series of any natural phenomena. Solar activity mainly outlines the solar variations.The main factors which outlines the solar variations are levels of solar radiation and ejection of solar material, the number and size of sunspots, solar flares, and coronal loops all exhibit a synchronized fluctuation, from active to quiet to active again, with a period of 11 years. Accumulation of solar nebula created the earth around 4.54 billion years ago. Volcanic outgassing probably created the primitive atmosphere, which contained almost no oxygen and would have been toxic to humans and most modern life. Much of the Earth was molten because of frequent collisions with other bodies. Over time, the planet cooled and formed a solid crust, eventually allowing liquid water to exist on the surface.

7.2 Solar Irradiance

Since 1978, solar irradiance has been directly measured by satellites, with very good accuracy. The Sun's total solar irradiance oscillates by +-0.1% over the ~11 years of the solar cycle, but that its average value has been stable since the measurements started in 1978. Solar irradiance before the 1970s used to be approximated using proxy variables, such as tree rings, the number of sunspots, and the abundances of cosmogenic isotopes such as ^{10}Be, all of which have been graduated to the post-1978 direct measurements.

7.3 Solar Cycle

Mean time period of Solar are about 11 years. Solar maximum and solar minimum are basically the repercussions of variations in the number of observed sunspots on the solar surface. Cycles span from one minimum to the next. Swiss astronomer Rudolf Wolf started numerating Solar Cycle on a practical basis. He started it from the year 1755 .Following Wolf's numbering scheme, the 1755–1766 cycle is traditionally numbered "1". Wolf created a standard sunspot number index, the Wolf index, which continues to be used today.

Factors Affecting the Solar Cycle of Solar Radiation Ejection

Levels of solar radiation and ejection of solar material have been judged by material in the solar wind which are also the interplanetary counterpart of coronal mass ejections (CMEs) called Interplanetary Coronal Mass Ejections or ICME's. Since early 1980s,the characteristic signatures of such materials, "interplanetary coronal mass ejections" or (ICMEs), had been identified. Limbs of Helios spacecraft and observations of the Solwind coronagraph demonstrated the clear association between CMEs at the Sun and shocks and ICMEs subsequently detected in the interplanetary medium. Solar wind plasma signatures of ICMEs include abnormally low proton temperatures, low electron temperatures, and bidirectional thermal electrons. Plasma compositional anomalies have also been identified in ICMEs, including enhanced plasma helium abundances relative to protons and occasional enhancements in minor ions (in particular iron) .Enhanced Fe charge states have also been reported. Such enrichments suggest that the plasma inside ICMEs originates in the low corona. Energetic particle signatures include bidirectional energetic protons and cosmic rays, energetic particle intensity depressions (Forbush decreases), and unusual solar energetic particle (SEP) flow directions.

7.4 Number and Size of Sunspots

The information about number and size of Sunspots are very necessary for understanding the mechanism behind the solar activity and solar cycle, and also for predicting the level of activity. Many other indicators of solar activity corresponds with Sunspot number very nicely . Sunspots are the excellent blueprints of solar magnetic flux and the area of a sunspot or a sunspot group has a better physical significance than sunspot number; an area of 130 msh (millionths of solar hemisphere; 1 msh ≈ 3×10^6 km^2) corresponds approximately to 10^{22} Mx (maxwell).

7.5 Solar Flares

A **solar flare** nothing but a sudden glaze of enhanced brightness on the Sun, usually observed near its surface and in immediate neighbourhood to a sunspot group. Powerful flares are often, escorted by a coronal mass ejection. Solar flares occur in a power-law spectrum of magnitudes; an energy release of typically 10^{20} joules of energy suffices to produce a clearly observable event, while a major event can emit up to 10^{25} joules.Solar flares often affect the solar cycle.

Coronal Loops

Coronal loops are huge loops of magnetic field beginning and ending on the Sun's photosphere projecting into the solar corona. They can be seen due to glazing ionized gas or plasma trapped in the loops. Coronal loops range widely in size up to several thousand kilometers long. They are momentary characteristics of the solar surface, forming and dissipating over periods of seconds to days. Coronal loops are related to sunspots since sunspots occur at regions of high magnetic field. The high magnetic field where the loop passes through the surface forms a barrier to convection currents, which bring hot plasma from the interior to the sun's surface, so the plasma in these high field regions is cooler than the rest of the sun's surface, appearing as a dark spot when viewed against the rest of the photosphere. The 11 year solar cycle depends upon the population of coronal loops which ,inturn affects the number of sunspots.

7.6 High Amplitude Cosmic Ray Intensity

The solar diurnal variation of cosmic-ray (CR) intensity shows a large day-to-day variability. This variability is a reflection of the continually changing conditions in the interplanetary space. The systematic and significant deviations in the amplitude/phase of the diurnal/semi-diurnal anisotropy from the average values are known to occur in association with strong geomagnetic activity. The enhanced diurnal variation of high-amplitude events exhibits a maximum intensity in space around the anti-garden hose direction and a minimum intensity around the garden hose direction.A number of high-amplitude events have been observed with a significant phase shift in the diurnal time of maximum to later hours or earlier hours.Such days are of particular significance when they occur during undisturbed solar/interplanetary conditions, as the superposed universal time effects are expected to be negligible. The diurnal variation might be influenced by the polarity of the magnetic field. The amplitude of the diurnal anisotropy for every individual HAE case is significantly larger than the quiet-day annual average amplitude and the phase of the diurnal anisotropy shifts to later hours for the majority of the events, as compared to the annual average values. Shocks in the interplanetary magnetic field have a capability to induce diffusion, convection and adiabatic energy losses plus curvature and gradient drift effects.

Magnetohydrodynamic shocks are the main inducors. Magnetohydrodynamic shocks leads towards shock acceleration. Magnetohydrodynamic Shock occurs through various mechanisms . Diffusive shock acceleration based on the first-order Fermi acceleration mechanism exactly adjudges the acceleration at the Earth's bow shock and traveling interplanetary shocks. Stochastic acceleration, or second-order Fermi acceleration is apparently the dominant mechanism in the vicinity of cometary bow shocks. CME-driven shocks moving outward from the Sun is also a mechanism. The increase in intensity enhancements leads to acceleration in interplanetary shocks augmented by major spectral changes. At energies $\lesssim 1$ MeV, the particles are accelerated out of the solar wind population by diffusive shock acceleration. For energies greater than 10 MeV, all observations to date of ionic charge states near the times of interplanetary shock arrivals are inconsistent with a solar wind origin, indicating that the ions were originally accelerated out of coronal material. The ions of ~ 1 MeV observed at the time of shock passage represent the same population as ambient solar energetic particles. Hence, from the above evidences ,it cen be elucidated that particles are injected into the interplanetary medium while the shock is still near the Sun, and later this particle population may be further affected by the shock as it propagates through the inner heliosphere.

Polarity of Solar Magnetic Field

A link exists between the cosmic ray intensity and the geomagnetic activity. A more elaborate the model. According to a detailed model, the modulated cosmic ray intensity measured by the ground based stations is equal to the galactic cosmic ray intensity (un-modulated) at a finite distance corrected by a few appropriate solar and terrestrial activity indices, causing the High Amplitude Wave Trains of Cosmic Ray Intensity. A strong geomagnetic activity is responsible for causing the structured and noteworthy deviations in the amplitude/phase of the high amplitude wave trains of cosmic ray intensity. The solar magnetic field polarity is dependent on the convective-diffusive mechanism, which relates the solar diurnal anisotropy of cosmic rays to the dynamics of the solar wind and of the interplanetary magnetic field. The field-aligned direction of the diffusive vector independently of the interplanetary magnetic field polarity has been confirmed. An investigation has been conducted by long-term changes in diurnal anisotropy of cosmic rays for the two solar cycles (20 and 21) during the period 1965–1990. They observed that the amplitude of the anisotropy is related to the characteristics of high and low amplitude days. The occurrence of high amplitude days are found to be positively correlated with the sunspot cycle while the low amplitude days are correlated negatively with the sunspot cycle. The high-speed solar wind streams (HSSWS) coming from coronal holes also leads to changes in the high amplitude wave trains of cosmic ray intensity . High Amplitude Wave Trains of Cosmic Ray Intensity are formed by two types of high-speed solar wind streams namely flare generated streams (FGS) and corotating streams (CS) .A deep correlation lies between the cosmic ray intensity decreases observed by high-speed streams produced by solar flares accompanied by Forbush decreases whose amplitudes are not directly correlated with the increase in solar wind speed. These latter decreases are usually large and are dependent on the location of the solar flares. Cosmic ray declinations related with coronal hole streams are much smaller than the typically Forbush-like depressions and no insubstantial difference is found in the Forbush-like decreases between the periods before and after the polarity changes.

7.6 Methodology and Data Analysis

In this study ,the Cosmic ray intensity data have been taken from the Bartol Neutron Monitor website (Website:www.bartol.udel.edu). Bartol Neutron Monitor is situated in the Department of Physics and Astronomy at the University of Delaware, England. It is housed in the H. Rodney Sharp Laboratory on the University campus. The primary function of the Institute is to carry out forefront scientific research, with a primary focus in physics, astronomy and space sciences. The 30 high amplitude events of cosmic ray intensity and 21 low amplitude events of cosmic ray intensity have been recorded and analysed , where 4 events each of high amplitude events and 3 events each of low amplitude events from the year 2008 to 2014 have been analyzed. The interplanetary magnetic field and solar wind velocity for each of the 30 and 21 events have been taken and recorded from the Omniweb Data Browser (www.omniweb.gsfc.nasa.gov). Omniweb Data Browser is primarily a 1963-to-current compilation of hourly-averaged, near-Earth solar wind magnetic field and plasma parameter data from several spacecraft in geocentric or L1 (Lagrange point) orbits. The data have been extensively cross compared, and, for some spacecraft and parameters, cross-normalized. Time-shifts of higher resolution data to expected magnetosphere-arrival times are done for data from spacecraft in L1 orbits (ISEE 3, Wind, ACE), prior to taking hourly averages from several IMP and GOES spacecraft. High Resolution OMNI (HRO) data set include 1-min and 5-min bow-shock-nose- shifted solar wind magnetic field and plasma data from IMP 8, Geotail, Wind and ACE spacecrafts.

7.7 Results

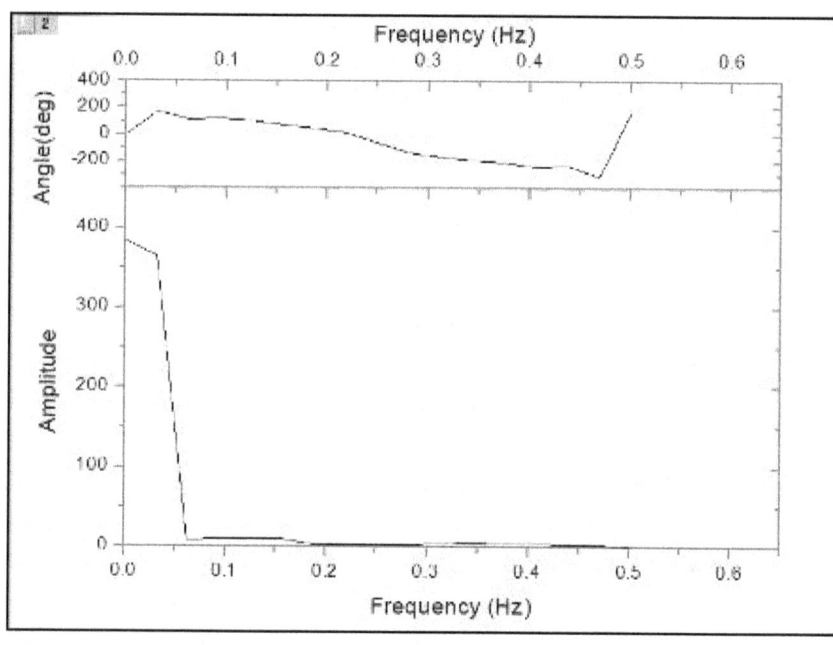

Fig 7.1 Amplitude Plot of High Amplitude Events of Cosmic Ray Intensity

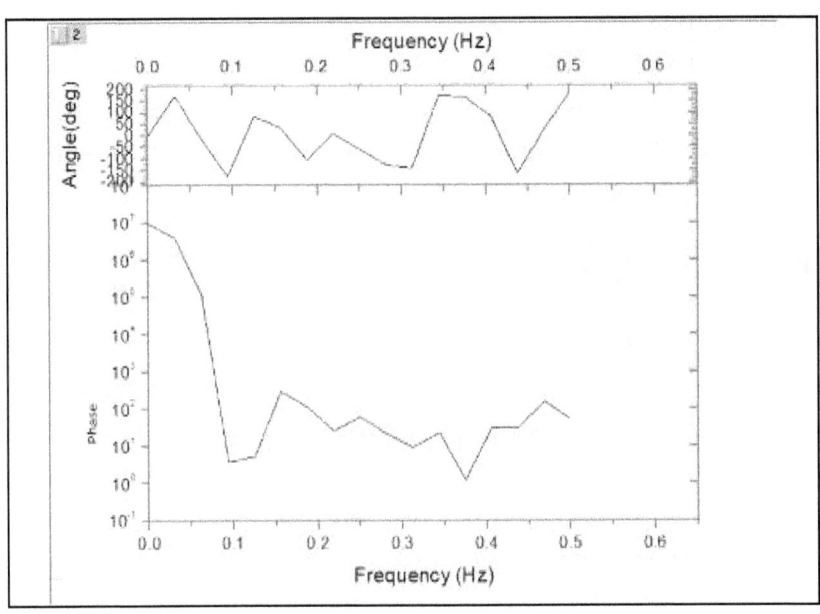

Figure 7.2 Phase Plot of High Amplitude Events of Cosmic Ray Intensity

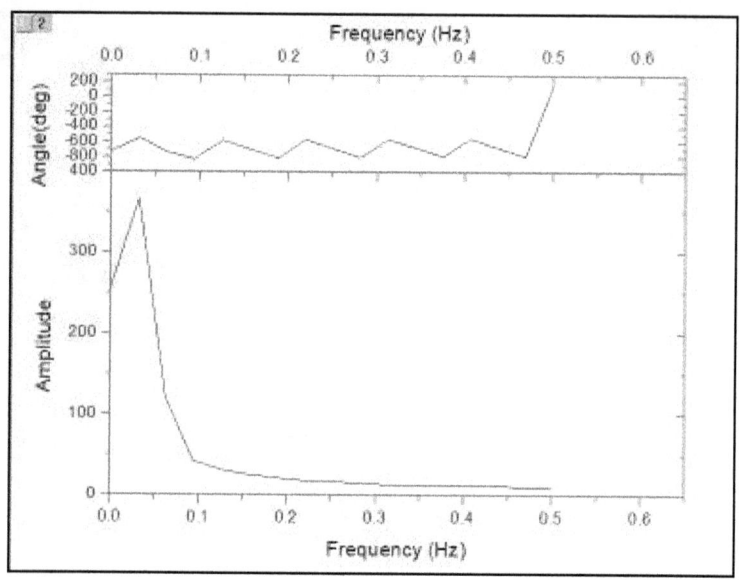

Fig 7.3 Amplitude Plot of Low Amplitude Events of Cosmic Ray Intensity

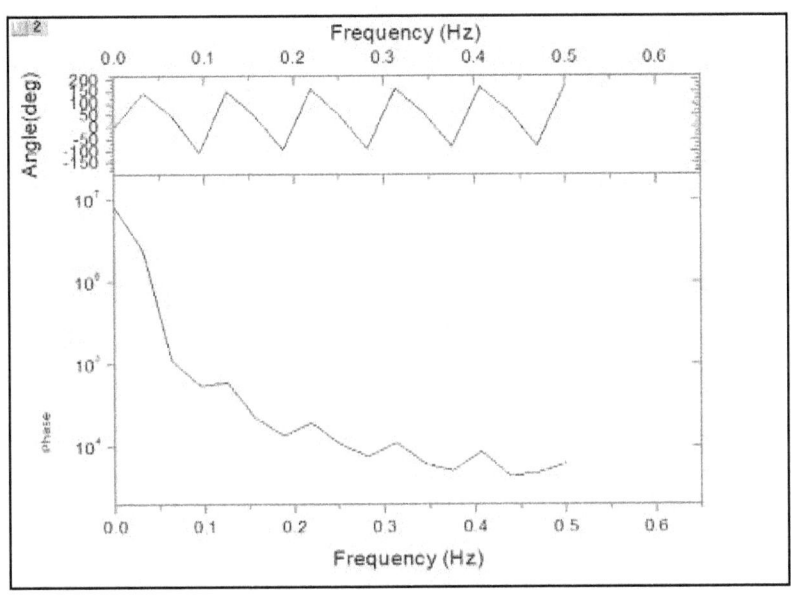

Fig 7.4 Phase Plot of Low Amplitude Events of Cosmic Ray Intensity

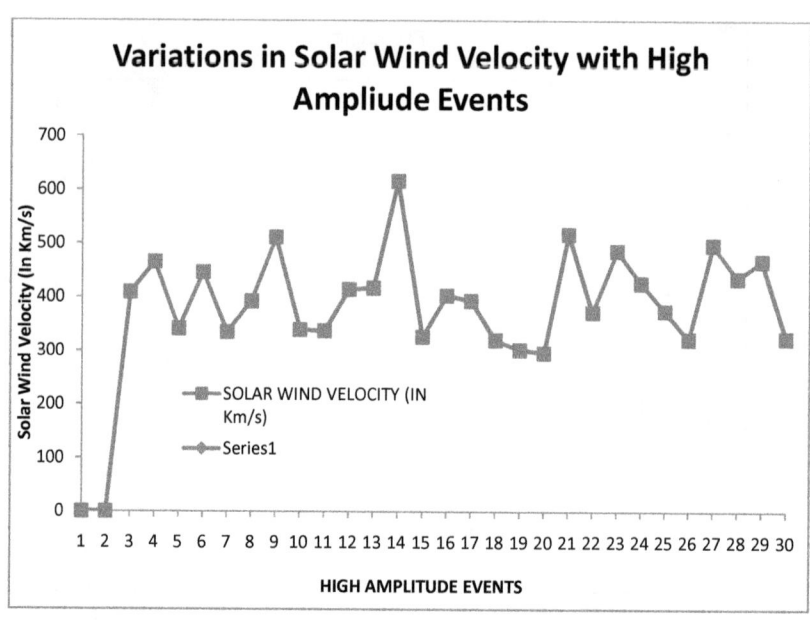

Fig 7.5 Variation in Solar Wind Velocity with CRI for High Amplitude Events

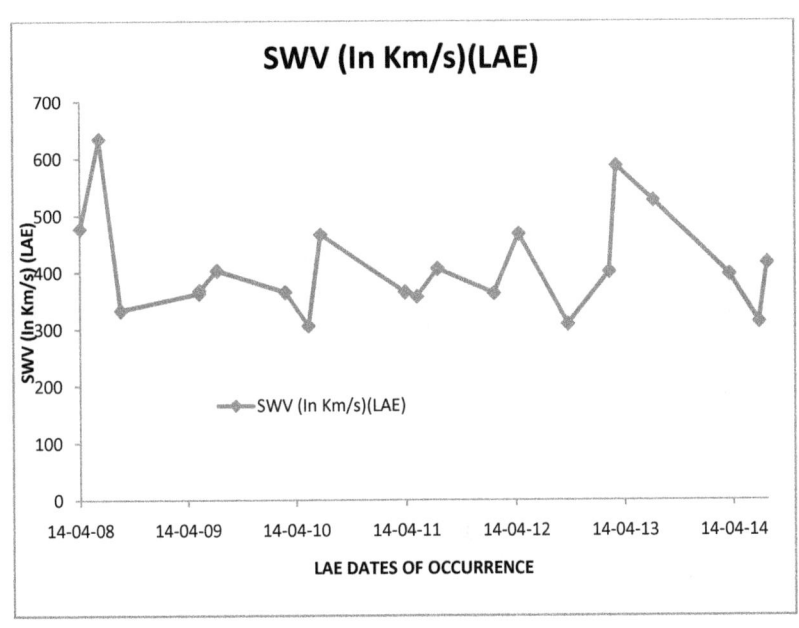

Fig 7.6 Variation in Solar Wind Velocity with CRI for Low Amplitude Events

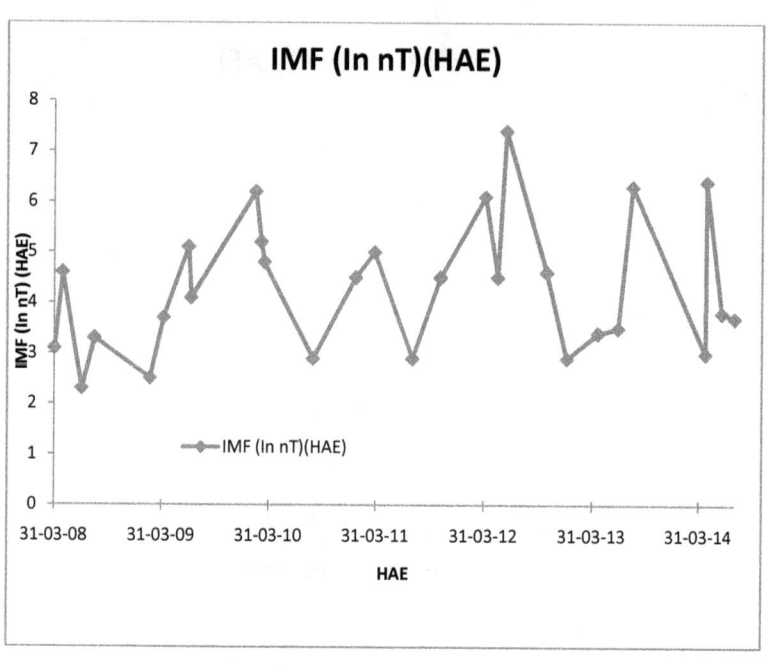

Figure 7.7 Variation in IMF with CRI for High Amplitude Events

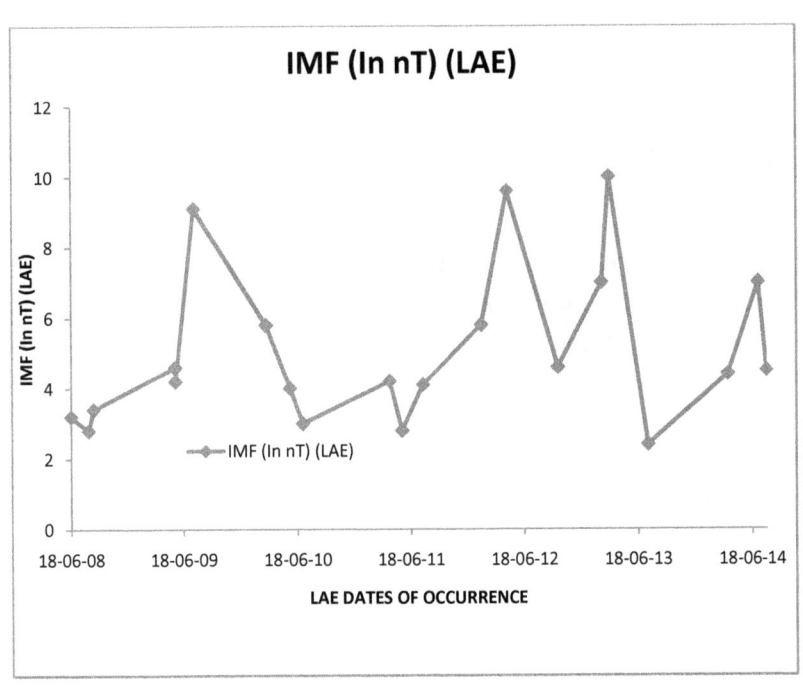

Fig 7.8 Variation in IMF with CRI for Low Amplitude Events

7.8 Discussion

Figure 7.1 shows the Amplitude Plot of Low Amplitude Events of Cosmic Ray Intensity values in the epoch 2008-2014. The plot shows a constant fall during the initial phase then it becomes constant between 0.1. and 0.2 hertz .Till 0.16 hertz ,it remains constant ,then goes on deceasing and becomes zero at 0.2 hertz .It remains zero from 0.2 to 0.3 hertz , then from 0.3 hertz ,it increases by a very trace amount and then remains constant .Then it reduces till 0.5 hertz ,at which it becomes zero. Figure 7.2 shows the Phase Plot of High Amplitude Events of Cosmic Ray Intensity values in the epoch 2008-2014.The phase value is the maximum at the first date of occurrence of High Amplitude Event , from where it goes on decreasing abruptly till 0.09 hertz ,from which it increases a bit till 0.14. hertz . Then it deceases suddenly till about 0.22 hertz ,then it increases by a trace amount till 0.25 hertz. Then it starts decreasing till 0.32 hertz ,again it increases by a trace amount till 0.34. hertz,then it deceases suddenly till about 0.37 hertz ,followed by abrupt increase and decrease till 0.5 hertz. Figure 7.3 shows the Amplitude Plot of Low Amplitude Events of Cosmic Ray Intensity values in the epoch 2008-2014.Initially , the value of amplitude is at 250 ,from which it experiences a steep rise till about 0.05 hertz frequency.After that , the frequency decreases till about 0.1 hertz ,followed by a gradual fall and constant values which goes on till 0.5 hertz.Hence ,the frequency values of amplitude plot follows a steep rise ,steep fall and constancy in the values of frequency. Figure 7.4 shows the Phase Plot of Low Amplitude Events of Cosmic Ray Intensity values in the epoch 2008-2014.Initially ,the phase is at 10^7 ,then it falls steeply to a value of 10^{63} ,then followed by a constant fall to a value of 10^5 ,then again a steep fall to a value of 10^{48}.After that ,a steep rise and fall occurs followed by constant fall which continues on till a phase values of 10^3 corresponding to a frequency value of 0.5 hertz . Figure 7.5 shows the variation of solar wind velocity with the high amplitude events during the epoch 2008-2014.In the year 2008 , the Solar wind velocity values follows a steep rise and fall accompanied by a steep rise. During the epoch 2009 , the interval between the first two events follows a steep fall accompanied by two types of rises between the 2nd and 3rd events- a slight steep rise and a constant rise and ultimately ,a steep fall between the 3rd and 4th events . During the year 2010 , values before the first event witnesses a small fall ,that between 1st and 2nd events shows a steep rise ,between 2nd and 3rd events exhibits a small rise again and ultimately between 3rd and 4th again a steep rise is witnessed. During the period 2012 , values before the first event witnesses a steep fall ,that between 1st and 2nd events shows a steep rise ,between 2nd and 3rd events exhibits a small fall again and ultimately between 3rd and 4th again a steep fall is witnessed. During the year 2013 , values before the first event witnesses a small fall ,that between 1st and 2nd events shows a small fall again ,between 2nd and 3rd events exhibits a steep rise is witnessed and between 3rd and 4th events again a steep rise is experienced . In the year 2014 , values before the first event witnesses a steep rise ,then accompanied by steep falls between the consecutive events. During the epoch 2011 , values before the first event witnesses a steep rise ,that between 1st and 2nd events shows a small fall ,between 2nd and 3rd events exhibits a small rise again and ultimately between 3rd and 4th again a steep fall is witnessed. Figure 7.6 shows the variation of solar wind velocity with the low amplitude events during the epoch 2008-2014.In the year 2008 , the Solar wind velocity values follows a steep rise and fall between the first ,second and third events respectively.During the epoch 2009 , the interval between the first two events witnesses a small rise ,between the 2nd and 3rd events- a steep rise is witnessed. During the year 2010 , the Solar wind velocity values follows a steep rise and fall between the first ,second and third events respectively .During the period 2012 , values between 1st and 2nd events shows a small. fall ,between 2nd and 3rd events exhibits a steep rise. During the year 2013 , the Solar wind velocity values follows a steep rise and fall between the first ,second and third events respectively. In the year 2014 ,interval between the 1st and 2nd events witnesses a steep rise followed by a steep fall between the 2nd and 3rd events .During the epoch 2011 , values between 1st and 2nd events shows a steep fall ,between 2nd and 3rd events exhibits a steep rise again.

Figure 7.7 shows the variation of interplanetary magnetic field with the high amplitude events during the epoch 2008-2014.In the year 2008 , the interplanetary magnetic field values follows a steep rise ,steep fall and again a steep rise between the 1st , 2nd , 3rd and 4th events respectively. During the epoch 2009 , the interval between the first two events follows a steep fall accompanied by two types of rises between the 2nd and 3rd events- a slight steep rise and a constant rise and ultimately ,a steep fall between the 3rd and 4th events. During the year 2010, values before the first event witnesses a small fall ,that between 1st and 2nd events shows a steep rise ,between 2nd and 3rd events exhibits a small rise again and ultimately between 3rd and 4th again a steep rise is witnessed. During the period 2012, values before the first event witnesses a steep fall ,that between 1st and 2nd events shows a steep rise ,between 2nd and 3rd events exhibits a small fall again and ultimately between 3rd and 4th again a steep fall is witnessed. During the year 2013, values before the first event witnesses a small fall ,that between 1st and 2nd events shows a small fall again ,between 2nd and 3rd events exhibits a steep rise is witnessed and between 3rd and 4th events again a steep rise is experienced. In the year 2014, values before the first event witnesses a steep rise ,then accompanied by steep falls between the consecutive events. During the epoch 2011, values before the first event witnesses a steep rise ,that between 1st and 2nd events shows a small fall ,between 2nd and 3rd events exhibits a small rise again and ultimately between 3rd and 4th again a steep fall is witnessed .Figure 7.8 shows the variation of interplanetary magnetic field with the low amplitude events during the epoch 2008-2014.In the year 2008, the interplanetary magnetic field values follows a steep rise and fall between the first ,second and third events respectively.During the epoch 2009, the interval between the first two events witnesses a small fall ,between the 2nd and 3rd events- a steep rise is witnessed. During the year 2010, the values follows a steep fall between the first ,second and third events respectively .During the period 2012 ,2013,2014 & 2011 respectively, values between 1st, 2nd and 3rd events shows a steep rise and fall. During the year 2013, the interplanetary magnetic field values follows a steep rise and fall between the first ,second and third events respectively.

7.9 Conclusion

There are some discontinuities in the tuning of solar wind velocity with cosmic ray intensity. The values of interplanetary magnetic field are in agreement with the rise and fall of the values of cosmic ray intensity. The level of cosmic ray flux which also signifies cosmic ray intensity arriving earth's lower atmosphere is an important parameter and has a causal relationship with the climate conditions of earth, and cloud formation.

Cosmic ray intensity was maximum during 2008, during last two values of 2009, during three values of 2011 and 2012.The annual rainfall during these years was measured to be the maximum .

CHAPTER 8
SUMMARY & CONCLUSION

In this last chapter deal the summary and conclusions of the research work after the detailed examination of results and their conclusions. Many assumptions and theories have been justified by this study. Firstly, Effect of solar activity on GCR modulation by taking 3 solar parameters *Sunspot numbers (SSN), Solar wind velocity and F10.7-cm solar radio flux and Interplanetary magnetic field (IMF)*.The variation of Sunspot number and Solar wind velocity with cosmic ray intensity on 13 high amplitude events and 7 low amplitude events during the low solar activity epoch ,i.e. from December 2008 to April 2014 have been discussed. The monthly averaged values of CR counts shows steep increase in the first quarter of 2015 and it continues on till 2017. In points where a steep rise in GCR intensity during the declining phase of cycle 24 is observed, GCR goes 6200 to 6800 counts in the December 2008 to March 2009 epoch. The variation of Sunspot number and F10.7 Solar flux unit with Kp ,Ap & DST indices on various CME events during the whole Solar Cycle 24 in many phases of solar cycle 24 were discussed in this Paper. In this study we have studied the wave analysis of amplitude and phase of the 21 Low Amplitude Cosmic ray events reaching earth.

Impact of Sunspot Number on Cosmic Ray Intensity

The monthly averaged values of CR counts shows steep increase in the first quarter of 2015 and it continues on till 2017. In points where a steep rise in GCR intensity during the declining phase of cycle 24 is observed ,GCR goes 6200 to 6800 counts in the December 2008 to March 2009 epoch. In the fig.2 , SSN values also shows a spotless trend during the descending epoch from December 2008 – April 2009 and October 2009 – December 2009.Observing above results the time variation of Sunspot Number during solar cycle 24 shows a very non-linear trend during the whole cycle. During the ascending epoch, duration January 2009-May 2009 and October 2009-December 2009 were spotless. There has been a steep rise and fall in SSN during the ascending and declining phases with the highest values of SSN achieved during the declining phase. In the duration May 2015-July 2015 and January 2016-March 2016 were also spotless . a significant increase in CRI was observed during the solar minima while decreasing at the peaks of the solar cycles. These trends reveal that with rise in SSN, the magnitude of heliospheric magnetic field rises and this inturn, decreases the inflow of galactic cosmic rays entering the solar system ,tending to a decrease in CRI and with fall in SSN,the inflow of GCR towards the solar system increases due to falling magnitude of heliospheric magnetic field ,leading to a decrease in CRI.

Impact of F10.7 solar radio flux on Cosmic Ray Modulation

The *time* variation of *F10.7 Solar Radio Flux* indicates that the initial phase of ascending epoch were spotless.After that ,the F10.7 solar radio flux remained almost constant from a value of 65 sfu to 75 sfu from January 2009 to January 2011 showing a simultaneous increase and decrease until January 2015.F10.7 solar radio flux values increases and decreases due to changing distances of earth from sun during earth's rotation and revolution. Hence variation of F10.7 solar radio flux with CRI is less pronounced.

Impact of Solar Wind Velocity on Cosmic Ray Intensity

In time variation of *solar wind velocity* during solar cycle 24 shows small fall in the initial stage of ascending epoch accompanied by a regular rise and fall during the ascending and declining phases of cycle. The highest value of *solar wind velocity* has been obtained in the declining epoch in the month of September 2017.The reason behind this is attributed to the beam of charged particles emitting from sun's corona escape from sun's gravitational field .The particles diversion occurs because they absorb the high temperature of sun's corona on moving into space and hence their velocity increases.Solar wind velocity reached highest value of 500 km/s during the solar maxima which is in agreement with high CRI values obtained during this period.

Impact of Interplanetary Magnetic Field on Cosmic Ray Modulation

The time variation of *Interplanetary Magnetic Field (IMF)* during solar cycle 24 shows a steep rise during the initial stage of the ascending epoch and followed regular rise during the whole cycle .Highest value of *Interplanetary Magnetic Field* was obtained in

January 2015 ,i.e. in the declining phase . The Regression Scatter Plots between Cosmic Ray Intensity and *Interplanetary Magnetic Field* indicates a negative correlation between them , the spectrogram not evenly distributed on two sides of the regression line .The IMF is immersed in the solar wind beam.IMF is proportional to SSN.In the descending phase of cycle 24,owing to smaller number of sunspots, the IMF was very low around 4 nT. Due to this descending phase, very small increase and decrease in IMF was observed. With the advent of ascending phase of cycle 24, SSN increased leading to an increase in IMF. This increase in IMF led to a decrease in CRI during this period evident by graphs.

Relation between Cosmic ray intensity and Solar Wind Velocity

The values of solar wind velocity show a linear decrease during the first three values, whereas the values of cosmic ray intensity show a simultaneous linear increase during the first three values. During the next two values, the solar wind velocity value shows a steep increase, whereas the cosmic ray intensity value shows a small linear increase. During the next two values, the solar wind velocity value shows a small steep decrease while the cosmic ray intensity values shows a corresponding steep increase. During the next two values, the solar wind velocity value shows a steep decrease while the cosmic ray intensity values show a small steep increase. The next two values of solar wind velocity witness a steep increase while the cosmic ray intensity values show a small increase. Last two values of solar wind velocity witness a steep decrease while the cosmic ray intensity value shows a small increase. From December 2008 to December 2009,solar minima was observed with much lesser number of sunspots, leading to smaller value of solar wind velocity. In December 2009, SSN abruptly rose to a value of 10 which attributed solar wind velocity a maximum value of 600 km/s. From January 2010 to April 2014, the SSN values fell rapidly followed by nominal increases and decreases due to which CRI values also varied linearly.

Relation between Cosmic Ray Intensity and Solar Wind Velocity

Above plot shows the time series of Solar Wind Velocity and Cosmic ray intensity with different low amplitude events during the descending phase of solar Cycle 24 ,i.e. from

December 2008-April 2014.The values of Solar Wind Velocity shows a steep increase during the first two values, whereas the values of cosmic ray intensity shows a small linear increase during the first two values. During the next two values ,the solar wind velocity values shows a steep decrease, whereas the cosmic ray intensity values shows a small linear increase. During the next two values ,the solar wind velocity values shows a small steep increase while the cosmic ray intensity values shows a corresponding small linear increase. During the next two values ,the solar wind velocity values shows a steep decrease while the cosmic ray intensity values shows a small linear increase. The next two values of solar wind velocity witnesses a steep increase while the cosmic ray intensity values shows a small linear increase. Last two values of solar wind velocity witnesses a small linear decrease while the cosmic ray intensity values also shows a corresponding linear decrease. Due to a constant smoothed SSN continuing till December 2009, the solar wind velocity exhibited a linear up and down curve because a constant quantity of charged particles escape from sun's gravity into space, but since SSN reduced to very low values from December 2009 to April 2014 accompanied by small increases time to time, hence linear curve of solar wind velocity is obtained also advocated by the CRI curves.

Relation between Cosmic Ray Intensity and SSN

The values of Sunspot Number show a simultaneous linear increase and decrease during the first three values, whereas the values of cosmic ray intensity shows a simultaneous small linear increase and decrease during the first three values. During the next two values ,the sunspot number values shows a minute steep increase, whereas the cosmic ray intensity values shows a small linear increase. During the next two values, the sunspot number values shows a small steep decrease while the cosmic ray intensity values shows a corresponding minute steep increase. During the next two values , the sunspot number values shows a steep decrease while the cosmic ray intensity values shows a small steep increase. The next two values of sunspot number witnesses a steep increase while the cosmic ray intensity value shows a small increase. Last two values of sunspot number witnesses a small increase while the cosmic ray intensity values shows constancy. The sunspot number was low from December 2008 to December 2009 due to low solar activity during that epoch. The curve was almost linear with periodic increase and decrease due to increasing SSN at some periods. The CRI values exhibited a linear graph.

Relation between Cosmic ray intensity and SSN

The values of Sunspot Number show constancy during the first two values, whereas the values of cosmic ray intensity show a small linear increase during the first two values. During the next two values, the sunspot number values show a small linear increase, whereas the cosmic ray intensity values shows corresponding constancy. During the next two values, the sunspot number values show a steep increase while the cosmic ray intensity values shows a corresponding small linear increase. During the next two values, the sunspot number value shows a steep decrease while the cosmic ray intensity values shows constancy. The next two values of sunspot number while the cosmic ray intensity values shows a corresponding constancy. Last two values of sunspot number witnesses a steep increase while the cosmic ray intensity values also show a corresponding small linear decrease. The SSN value shows a linear plot with a steep increase and decrease in the middle of cycle 24.The CRI values were linear during this time.

Variation between Solar Activity and Earth's Upper Atmospheric Parameters during Coronal Mass Ejections

A **Coronal Mass Ejection (CME)** is a remarkable ejection of plasma and escorting magnetic field from the solar corona. They often go after solar flares and are usually present during a solar prominence eruption.CMEs are mainly produced from active regions on the Sun's surface, like clusters of sunspots associated with frequent flares. Near solar maxima, the Sun produces about three CMEs every day, whereas near solar minima, there is about one CME every five days.From the study of variations between solar activity and Earth's Upper Atmospheric Parameters observed during 51 special CME events during the whole epoch of cycle 24, the results obtained are as follows:

Variation between F10.7 index and Kp Index

For this 11-year interval, the spectra of both parameters have a vertical offset showing uniform variation during the February 2010 to March 2011 epoch. This constancy is due to the fact that cycle 24 was minimum during the period February 2010 to March 2011 with negligible number of sunspots being tracked at that time. Sunspot values indicate that solar maxima had its advent in March 2012. Both F10.7 and Kp index spectra show a peak during the periods of April 2011 to August 2011 ,January 2012 to August 2012, October 2012 to December 2012 and October 2013 to December 2013 respectively with an exception for the period of April 2013 to October 2013 where both spectra shows a negative relationship. Both spectra showed disagreement during cycles 21, 22 and 23 . Positive relationship among both spectra during February 2010-March 2011 period is due to highly lesser number of sunspots during this time. Lesser number of sunspots leads to lesser number of CME's exhibiting a straight F10.7 and Kp index line on the plot. On July 2011 owing to more number of sunspots, the Kp index witnessed a small increase followed by a decrease. January 2012 to August 2012 was marked by high values of F10.7 Index of 158 sfu and Kp Index as 40 respectively on June 2012 and March 2012 .The horizontal component of earth's magnetic field is interrupted in March 2012 by the eruption of large number of active flares with some prominent flares being earth-directed striking high energy electromagnetic rays on Earth in minutes. This period was characterized by some strong and moderate geomagnetic storms. On June 2012 , NASA announced the highest flux of gamma rays having magnitude greater than 100 MeV related to an eruption on the Sun creating a peak in F10.7 index in June 2012[NASA's Official Website]. The unexpectedly low solar activity continued in April 2013 till April 2014. Nevertheless solar minima continued till April 2014, then also four 13 M-class flares, directed X shatters were observed causing Kp index to achieve its peak on June 2013.These flares generated a strong R3 radio blackout in the upper atmosphere.

Variation between F10.7 index and Dst Index

Spectra of both parameters exhibits a linear relationship during the period of February 2010 to March 2011 owing to a very small magnitude of Sunspot number during this time F10.7 and Kp index respectively reached their highest values of 180 sfu and -20 nT on

April 2011.*Both* spectra exhibited a negative correlation at this epoch. This negative correlation came due to solar minima which started since the beginning of Cycle 24 and continued on till early 2014 preceded by solar flares ,sunspots and "double-peaked" solar maximum. From April 2011 to August 2011, solar minima causes a fall in F10.7 index corresponding to which Dst index values approached zero. This unusual behavior is due to lower solar activity leading to lesser amount of solar flares, solar winds , and other solar ejections directed towards earth .This strengthened the earth's magnetic field. The charged particles directed towards earth due to solar emissions generated a magnetic field against the Earth's magnetic field and ultimately, areal magnetic field reduced. The low solar activity caused Dst index to achieve its ever possible minimum value of approximately -80 nT in March 2012.

Variation between F10.7 Index and Ap Index

Time Variation between F10.7 index and Ap index follows the same trend as Figure 1 except during April 2013 to October 2013 period when there was a simultaneous increase and decrease in both parameters. The solar activity was very abnormally minimal in April due to which F10.7 index fell to 100 sfu. Abrupt increase in solar activity in May 2013 led F10.7 index to increase leading to expulsions of many solar flares .Since these flares were not geoeffective, hence Dst index fell rapidly till October 2013.April 2013 to October 2013 was a period of negligible striking of charged particles on earth, hence earth's magnetic field was very strong .This trend has been the reverse as observed in Solar cycle 23 where increase in F10.7 index led to a decrease in Ap index.

Variation between Sunspot Number and Kp Index

In results shows the time series analysis of Sunspot Number and Kp index for 51 CME's for cycle 24 .Both spectra show linear relationship during every period except for the period of September 2011 to January 2012 during which both exhibit a negative correlation. Cycle 24 had the highest number of CME driven storms as compared to preceding solar cycles[18].The F10.7 index increased till January 2012 due to continuous increase in the Sunspot number with the highest Sunspot number value being registered as 130 in November 2011 due to which Kp index surged to a value of 110. During the period of October 2013 to December 2013, spectra of both parameters show a peculiar relationship due to abnormally low solar activity recording the smallest fall in sunspot number .Large number of solar flares erupted during this time, but majority of them were not earth-directed, hence Kp index fall rapidly up to November 2013, after which some prominent earth-directed solar flares erupted reporting a surge in Kp index values.

Variation between Sunspot Number and Dst Index

In the results shows the time series analysis of Sunspot Number and Dst index for 51 CME's for cycle 24 .Both spectra are positively correlated during the whole period of cycle 24 with an exception during the period of August 2011 to August 2012.The number of sunspots were restricted to one or two during this period ,although ,a large number of flares erupted which were not geoeffective due to which no or very few charged particles came towards the earth and interacted with its magnetosphere ,hence an abrupt recession in Dst values were reported till March 2012 was observed .In the time span of March 2012 to August 2012 ,some major flares which were directed towards earth occurred like an M-class flare which is a very active flare causing Dst index to rise and reach its maximum value on May 2012 due to a very huge sunspot termed monster sunspot emitting a huge flare the M5.7 flare into space, equivalent to the dimensions of Jupiter, or eleven times the diameter of Earth.

Variation between Sunspot Number and Ap Index

Results Figure 6 shows the time series analysis of Sunspot Number and Ap index for 51 CME's for cycle 24 .Spectra of both parameters show a strictly linear relationship except during the period of May 2013 to November 2013 with rise in sunspot number accompanied by fall in Ap index ,reaching to their apex values on June 2013 and again following the same negative correlation. Till May 2013 , the solar activity was abruptly minimal due to which sunspot number fell drastically to the lowest ever value of 25.Solar activity rose very quickly in May 2013 reporting a very large linear increase in sunspot number with highest sunspot number being communicated as 150 and small undeviating increase in Ap index with its value noted as 20 nT. From June 2013 to November 2013, many flares and prominences arose which were not geoeffective .hence geomagnetic activity fell intensely hence a linear fall in Ap index was registered during this time.

Scatter Plot between F10.7 index and Kp Index

Using data from Omni web (Website: https://omniweb.gsfc.nasa.gov/html/ow_data.html) which is a website of Goddard Space Flight Centre ,NASA ,there is a weak positive correlation between F10.7 index and Kp index .A very small value of R = 0.216 indicates an increase in Kp index values with increase in F10.7 index values but this variation is not pronounced as the data points are not tighter along the trendline.It is also observed in the Time Series Analysis between the two parameters.This weak correlation is due to the fact that many flares arose during this period ,predominantly earth- directed .Sunspot number also exhibited an exponential rise and fall during this time .Both parameters have not showed a strict linear correlation .

Scatter Plot between F10.7 index and Dst Index

A feeble value of R=0.1 indicates non-linear relationship between the two variables. The time series graph also advocated the fact that change in F10.7 index have not rigorously led to change in Dst index values. The monster sunspot erupted releasing a very predominant earth-directed solar flare into space. Negative Dst values were prevalent during this period, due to which earth's magnetic field got weakened.

Scatter Plot between F10.7 index and Ap Index

A very small value of R = 0.021 points towards a weak positive correlation between the two variables .The period of observation of the analysis has witnessed merely 40% of the geomagnetic activities, leading to this weak correlation.

Scatter Plot between Sunspot Number and Kp Index

On plotting scatter plots between sunspot number and Kp index, a very weak correlation coefficient "(R = 0.07)" was observed. This was due to solar minima continuing throughout the entire period of observation due to which sunspot number was predominantly low. This led Kp index towards an abnormal variation.

Scatter Plot between Sunspot Number and Dst Index

Scatter plots between Sunspot number and Dst index exhibits a very strong correlation (R = 0.32). An upward moving trend line with tighter data points falling along the trendline was observed .Reason behind this strong correlation has been a constant unexpectedly low solar activity during the period of observation. This led to a constant sunspot number value neither low nor high, hence the Dst index values also varied equally with sunspot number variations.

Scatter Plot between Sunspot Number and Ap Index

A very poor value of correlation coefficient "(R = 0.14)" indicates a very weak positive correlation between sunspot number and Ap index .Since Ap index indicates the average value of geomagnetic storms ,low value of correlation coefficient shows abnormal change in the magnitude of geomagnetic storm values with increase in sunspot number values.

Wave Analysis of Cosmic rays and Solar Wind and Geomagnetic Index (Ap)

The Cosmic Ray Intensity values of 3 major Low Amplitude events have been taken from 2011 to 2017 and the Fast Fourier Transform have been plotted. The Cosmic ray intensity data have been taken from the Bartol Research Institute's Neutron Monitor Station situated in the University of Delaware.(Website:neutron.bartol.udel.edu).The variation of CRI values with Ap index and Solar Wind during these 3 major Low Amplitude events have been studied. The Amplitude Plot shows an abrupt fall in the year 2011 followed by steep rise and fall from years 2012 to 2017. Phase plot of Fast Fourier Transform of the epoch 2011-2017 shows a steep decrease during the year 2011 followed by small increases and decreases during the whole epoch.

Variation in Cosmic Ray Intensity with Ap Index

Variation in Ap index with cosmic ray intensity.Both parameters show simultaneous variation in epochs 2013, 2016 ,2014 and 2015 and non-uniform variation in epochs 2011, 2012 and 2017. The geomagnetic activity index Ap on an average basis remains low during the period of each HAE/LAE the cases have been investigated with an exception in the year 2017 ,when it reached a high value of 526 ,due to sudden increase in the number of sunspots of that epoch whose data has been taken as till that time , there were no sunspots. This is because the interplanetary disturbances responsible for cosmic-ray modulation effects have not reached the Earth till that period of abrupt increase since prior to that, there were no sunspots. Further, the solar activity on the back side is not likely to produce the usual terrestrial manifestations such as geomagnetic storms, produced by activity on the visible side of the Sun. However, modulation by plasma clouds ejected from the back side and the subsequent propagation of cosmic-ray particles might reveal such flare activity. If this is so, the HAE/LAE in the absence of geomagnetic storm.This is because the solar activity was high during periods 2013, 2014, 2015 and 2016 and was low during periods 2011, 2012 and 2017due to which geomagnetic activity increased and cosmic ray intensity increased with increase in Ap index,except at some points where the sunspot number was maximum.

Variation in Cosmic Ray Intensity with Solar wind velocity

Above plot shows the variation in Solar Wind Velocity with Cosmic ray intensity. The Cosmic Ray intensity shows an uniform variation during the whole epoch whereas Solar Wind Velocity exhibit variation in the whole epoch. Majority of large amplitude events occur at an average solar wind velocity. This distribution is quite broader ranging from low to near high solar wind velocity. Hence it can be inferred that large amplitude events are less or weakly dependent on the solar wind velocity. The solar wind velocity decreased due to low solar activity. There were very less sunspots, hence less solar wind speed and lesser number of sunspots. This led to constant CRI. During April 2014, when solar activity became maximum, the CRI value increased. CRI remained constant because of the interplanetary magnetic field.

Variation of Cosmic Ray Intensity with Solar Wind Velocity

In the year 2008, the Solar wind velocity values follows a steep rise and fall accompanied by a steep rise. During the epoch 2009 , the interval between the first two events follows a steep fall accompanied by two types of rises between the 2nd and 3rd events- a slight steep rise and a constant rise and ultimately ,a steep fall between the 3rd and 4th events . During the year 2010 , values before the first event witnesses a small fall, that between 1st and 2nd events shows a steep rise ,between 2nd and 3rd events exhibits a small rise again and ultimately between 3rd and 4th again a steep rise is witnessed. During the period 2012 , values before the first event witnesses a steep fall ,that between 1st and 2nd events shows a steep rise ,between 2nd and 3rd events exhibits a small fall again and ultimately between 3rd and 4th again a steep fall is witnessed. During the year 2013 , values before the first event witnesses a small fall ,that between 1st and 2nd events shows a small fall again ,between 2nd and 3rd events exhibits a steep rise is witnessed and between 3rd and 4th events again a steep rise is experienced . In the year 2014 , values before the first event witnesses a steep rise ,then accompanied by steep falls between the consecutive events.during the epoch 2011 , values before the first event witnesses a steep rise ,that between 1st and 2nd events shows a small fall ,between 2nd and 3rd events exhibits a small rise again and ultimately between 3rd and 4th again a steep fall is witnessed .The solar wind velocity exhibited a steep increase and decrease due to low solar activity till the end of 2009.The solar wind velocity started increasing from 2009 to 2010,because rise in solar activity after 2009 led to rise in solar magnetic field, hence increase in corona's temperature which continued on till mid of 2014.The solar wind velocity achieved its highest value of 650 km/s during this period. A steep increase and decrease in solar wind velocity continued till the end of cycle 24 which was evidenced by some sunspots giving birth to crucial earth-directed CME's.

Variation in Cosmic Ray Intensity with Solar Wind Velocity

In the year 2008, the Solar wind velocity values follows a steep rise and fall between the first ,second and third events respectively. During the epoch 2009, the interval between the first two events witnesses a small rise, between the 2nd and 3rd events- a steep rise is witnessed. During the year 2010, the Solar wind velocity values follows a steep rise and fall between the first, second and third events respectively. During the period 2012, values between 1st and 2nd events shows a small fall, between 2nd and 3rd events exhibits a steep rise. During the year 2013, the Solar wind velocity values follows a steep rise and fall between the first, second and third events respectively. In the year 2014, interval between the 1st and 2nd events witnesses a steep rise followed by a steep fall between the 2nd and 3rd events. During the epoch 2011, values between 1st and 2nd events shows a steep fall, between 2nd and 3rd events exhibits a steep rise again. The solar wind velocity increased to the maximum value of 650 km/s due to some earth directed CME's which appeared during mid-2008 and again decreased to its lowest value during the end of 2008. Due to low solar activity from 2008 to mid-2011, solar wind velocity remained minimal and exhibited a small increase and decrease. This continued on till 2013, when it again reached its maximum value due to occurrence of earth-directed CME's.

Impact of Interplanetary Magnetic Field on Cosmic Ray Intensity

The interplanetary magnetic field is immersed in the solar wind beam. It is again proportional to the sunspot number. In the descending phase of cycle 24, due to smaller number of sunspots, the interplanetary magnetic field was a very low value of 4 nT. Due to this descending phase, very small increase and decrease in IMF was observed. With the advent of ascending phase of cycle 24 in June 2014, SSN increased leading to an increase in IMF. This increase in IMF led to an increase in CRI during this period evident by the time series graphs.

Variation in Cosmic Ray Intensity with Interplanetary Magnetic Field

In the year 2008, the interplanetary magnetic field values follows a steep rise, steep fall and again a steep rise between the 1st, 2nd, 3rd and 4th events respectively. During the epoch 2009, the interval between the first two events follows a steep fall accompanied by two types of rises between the 2nd and 3rd events- a slight steep rise and a constant rise and ultimately, a steep fall between the 3rd and 4th events. During the year 2010, values before the first event witnesses a small fall, that between 1st and 2nd events shows a steep rise, between 2nd and 3rd events exhibits a small rise again and ultimately between 3rd and 4th again a steep rise is witnessed. During the period 2012, values before the first event witnesses a steep fall ,that between 1st and 2nd events shows a steep rise ,between 2nd and 3rd events exhibits a small fall again and ultimately between 3rd and 4th again a steep fall is witnessed. During the year 2013 , values before the first event witnesses a small fall ,that between 1st and 2nd events shows a small fall again ,between 2nd and 3rd events exhibits a steep rise is witnessed and between 3rd and 4th events again a steep rise is experienced. In the year 2014, values before the first event witnesses a steep rise ,then accompanied by steep falls between the consecutive events. During the epoch 2011 , values before the first event witnesses a steep rise ,that between 1st and 2nd events shows a small fall ,between 2nd and 3rd events exhibits a small rise again and ultimately between 3rd and 4th again a steep fall is witnessed .The IMF exhibited a small increase and decrease except during the initial period of 2008,mid-2010 to mid-2012 and mid-2013 to mid-2014.In the first half of 2008,two major sunspots 1007 and 1009 appeared causing intense solar flares. These flares increased the interplanetary magnetic field. In the period mid 2010 to mid-2012, sunspots 1041 and 1046 and sunspot groups 1121,1164 and 1166 appeared, among which sunspot group 1164 caused a CME of speed 2200 km/s directed towards earth. This CME increased the value of IMF and an IMF of 7.5 nT was obtained. Although solar activity was low, but 72.4 sunspots were observed, hence IMF was high.

Variation in Cosmic Ray Intensity with Interplanetary Magnetic Field

In the year 2008, the interplanetary magnetic field values follows a steep rise and fall between the first, second and third events respectively. During the epoch 2009, the interval between the first two events witnesses a small fall, between the 2nd and 3rd events a steep rise is witnessed. During the year 2010 , the values follows a steep fall between the first, second and third events respectively .During the period 2012 ,2013,2014 & 2011 respectively, values between 1st , 2nd and 3rd events shows a steep rise and fall. During the year 2013, the interplanetary magnetic field values follows a steep rise and fall between the first, second and third events respectively. The IMF was very low in 2008, due to extremely low solar activity.IMF decreased from 2008 to mid-2009 which emerged leading to large number of CME's. Then IMF decreased from mid-2009 to the beginning of 2010, and then it followed a steep increase and decrease, followed by some exceptions during mid-2012 and middle 2013 during which it achieved a highest value of 10.5 nt, due to eruption of active regions.

Future Scope of the Work

The research work carried out by me has a very wide future scope. In my first objective, I came to the conclusion that mainly SSN and F10.7 Solar radio flux are responsible for cosmic ray modulation. Among various indicators of solar activity, SSN is the main parameter which leads to changes in CRI and implications for the near future is that a modeling between SSN and CRI can be definitely exploited for most effective convergence towards best solution of space weather studies. A study of a relation between SSN and Dst index will definitely go a long way to mitigate effects of climate changes which we are leaving to future research scholars for further research as it was beyond the scope of our study. We adopted Fourier analysis to decompose CRI values into Amplitude and Phase waves. We came to the conclusion that Amplitude wave is a triangular wave and Phase wave is a plane wave. Wave analysis can be further done to assess the harmful effects of cosmic rays on satellites, power grids etc. We also came to a conclusion that interplanetary magnetic field are in tune with the rise and fall of the CRI values and cosmic ray flux level signifying CRI arriving earth's lower atmosphere is an important parameter and has a causal relationship with the climate conditions of earth , and cloud formation ,hence further research can be carried out to develop a model which can be accepted globally establishing a relationship between IMF , CRI and climate change taking into account the climatic conditions of earth and cloud formation. It will help in weather forecasting on earth and helping to mitigate climate change.

CPSIA information can be obtained
at www.ICGtesting.com
Printed in the USA
LVHW051938140723
752119LV00012B/671